Neogene Avian Localities of North America

Neogene Avian Localities of North America

JONATHAN J. BECKER

SMITHSONIAN INSTITUTION PRESS
Washington, D.C.
London

Copyright © 1987 by Smithsonian Institution
All rights reserved

LIBRARY OF CONGRESS CATALOGING-IN-PUBLICATION DATA

Becker, Jonathan J., 1955–
Neogene avian localities of North America.

Bibliography: p.
Includes indexes.
1. Birds, Fossil. 2. Paleontology—Tertiary.
3. Paleontology—North America. I. Title.
QE871.B346 1987   568'.097   87-42557
ISBN 0-87474-225-0 (pbk.)

Contents

Abstract   1
Introduction   3
Acknowledgments   7
Checklist of Neogene Birds of North America   9
North American Neogene Avian Localities
   Late Arikareean   25
   Hemingfordian   27
   Barstovian   36
   Clarendonian   46
   Hemphillian   63
   Blancan   87
Literature Cited   109
Index   147
Figures   167

Abstract

This is a synopsis of all North American Neogene (Miocene and Pliocene) localities that have produced fossil birds. Widely scattered literature is compiled and corrected and each locality is presented with current information as to the geologic formation, political state, relative age based on the stage of evolution of the included land mammals, absolute dates if available, and a list of avian taxa present. Preliminary identifications of unstudied avifaunas are given. A bibliography is provided for the avian taxa, and includes if available, pertinent references to the geological age and the associated vertebrate fauna of each locality.

The distribution of avian taxa and fossil localities are sharply skewed in the geological record of North America. The last half of the Neogene (Clarendonian through Blancan) is much better sampled that the first half (late Arikareean through Barstovian). The diversity of the fossil taxa of birds roughly parallels the distribution of localities.

Introduction

This report is a synopsis of the Neogene localities in North America that have produced fossil birds, arranged by North American Land Mammal Age. Each entry includes the name of the fauna or local fauna, the geologic formation, the political state, the relative age based on the stage of evolution of the land mammals or marine correlative, the absolute dates if available, and a list of the described avian taxa. A bibliography of these avian taxa, and a few pertinent, general references to the geologic age or associated vertebrate fauna of each locality are included. This report includes all published localities in North America. Most localities are from the United States, but a few are from southern Canada and northern Mexico. I am unaware of any Neogene locality that contains birds from Mexico through Panama, or from northern Canada. Species marked with an asterisk (*) indicate it as being from the type-locality.

The purpose of this list is to correct the widely scattered reports in the literature of the Neogene birds and to compile them faunistically, with information on the correct geologic age and references to the associated fauna. I have also added preliminary identifications for many of the unstudied avifaunas. I hope that this report will provide the impetus for taxonomic revision of questionable records, correction of erroneous reports, and study of additional material. This abstract allows us to view Neogene avian evolution faunistically rather than systematically and also provides a correlated list of avian taxa to facilitate paleoecological studies.

The Neogene includes the Miocene and Pliocene epochs. The Miocene Epoch spans the time interval from about 22.5 million years ago to slightly less than 5 million years ago. (See comments by Savage and Russell [1983] and

Berggren and Van Couvering [1974] concerning the boundaries of the Miocene). In North America, the Miocene Epoch has been divided into the following Land Mammal Ages: late Arikareean (22.5 - 20 m.y.), Hemingfordian (20 - 16 m.y.), Barstovian (16 - 11.5 m.y.), Clarendonian (11.5 - 9 m.y.) and Hemphillian (9 - 4.5 m.y.). Prior to the redefinition of the Miocene Epoch to include the Messinian Age, the Clarendonian and Hemphillian Land Mammal Ages had been termed the Early and Middle Pliocene, respectively. The Pliocene is considered here to be essentially synonymous with the Blancan Land Mammal Age (4.5 - 1.8 m.y.).

Correlations (Figure 1) are primarily based on the works of Berggren and Van Couvering (1974), Repenning (1977), Berggren et al. (1985), and especially Tedford et al. (in press). Radiometric dates are mainly from Tedford et al. (in press).

The general arrangement of fossil birds follows Brodkorb's Catalogue of Fossil Birds (1963c, 1964b, 1967, 1971a, 1978) with the following major exceptions: Palaelodidae included in the Phoenicopteridae (Olson and Feduccia 1980); Phoenicopteridae moved to Charadriiformes (Olson and Feduccia 1980): the genera Nettion and Querquedula included in Anas (Johnsgard 1979); Pseudodontornithidae and Cyphornithidae included in Pelagornithidae (Olson 1985b); Aramornis transferred from Aramidae to Balearicinae (Gruidae) (Olson 1985b), Paractiornis transferred from Haematopodidae to Glareolidae (Olson and Steadman 1978); Palostralegus synonymized with Haematopus (Olson and Steadman 1978); Rhegminornis transferred from Jacanidae to Meleagridinae (Olson and Farrand 1974); and Palaeostruthus synonymized with Ammodramus (Steadman 1981).

There are still many unstudied Neogene avifaunas in North America that would affect any systematic, biogeographic, or biochronologic study based solely on the material herein. The published record of Neogene birds is a mosaic of well over one hundred years of differing systematic philosophies and techniques. Therefore, it is likely that some species included here are not valid and others may prove to have been incorrectly assigned to genus, family, or order.

The distribution of avian taxa and fossil localities are sharply skewed in the geological record of North America. The last half of the Neogene (Clarendonian through Blancan) is much better sampled that the first half (late Arikareean through Barstovian) (Figures 2, 3). The diversity of the fossil taxa of birds roughly parallels the distribution of localities.

I do not discuss or revise the systematic position of fossil taxa included herein. Olson's (1985b) recent review of the fossil record of birds should be consulted for these aspects of the North American avifauna.

In offering this synopsis I realize that other researchers will be aware of Neogene localities with birds and of references that I have overlooked. I hope they will draw my attention to this information so I may updating this list. I would also appreciate receiving notice of new papers that pertaining either to the fossil localities included here or to new localities as they may be discovered.

Acknowledgments

I thank the following people for their comments on this manuscript and for freely sharing unpublished information with me: L. G. Barnes, K. J. Bickart, P. Brodkorb, R. Chandler, K. Warheit, H. Howard, B. J. MacFadden, G. S. Morgan, C. Mourer-Chauvire, D. W. Steadman, S. D. Webb, and especially S. L. Olson and R. H. Tedford.

This manuscript was written while at the Department of Zoology, University of Florida and at the National Museum of Natural History, Smithsonian Institution. I thank G. Kiltie (University of Florida) for typing early drafts of this manuscript. Financial support received, includes in part, grants from the Frank M. Chapman Memorial Fund, American Museum of Natural History, and from Sigma Xi Grants-In-Aid of research, and a Smithsonian Postdoctoral Fellowship.

Checklist of Neogene Birds of North America

Order Gaviiformes Wetmore and W. D. Miller.
  Family Gaviidae Allen. Loons.
    Genus Gavia Forster
      G. brodkorbi Howard 1978
      G. concinna Wetmore 1940
      G. howardae Brodkorb 1953
      G. palaeodytes Wetmore 1943
Order Podicipediformes (Fürbringer).
  Family Podicipedidae (Bonaparte). Grebes.
    Genus Rollandia Bonaparte
      Rollandia sp.
    Genus Tachybaptus Reichenbach
      Tachybaptus sp.
    Genus Pliolymbus Murray
      P. baryosteus Murray 1967
    Genus Podiceps Latham
      P. discors Murray 1967
      P. parvus (Shufeldt 1913)
      P. subparvus (L. Miller and Bowman 1958)
      P. nigricollis Brehm 1831
    Genus Aechmophorus Coues
      A. elasson Murray 1967
    Genus Pliodytes Brodkorb
      P. lanquisti Brodkorb 1953
    Genus Podilymbus Lesson
      P. majusculus Murray 1967
      P. podiceps (Linnaeus 1758)
Order Procellariiformes Fürbringer.
  Family Diomedeidae. Albatrosses.
    Genus Diomedea Linnaeus
      D. cf. D. anglica Lydekker 1891
      D. californica Miller 1962
      D. milleri Howard 1966

Family Procellariidae (Boie). **Shearwaters and Allies.**
  Genus *Fulmarus* Linnaeus
    F. hammeri Howard 1968
    F. miocaenus Howard 1984
  Genus *Puffinus* Brisson
    P. barnesi Howard 1978
    P. calhouni Howard 1968
    P. conradi Marsh 1870
    P. diatomicus L. Miller 1925
    P. felthami Howard 1949
    P. inceptor Wetmore 1930
    P. kanakoffi Howard 1949
    P. micraulax Brodkorb 1963
    P. mitchelli L. Miller 1961
    P. priscus L. Miller 1961
    P. tedfordi Howard 1971
Family Oceanitidae (Forbes). **Storm Petrels.**
  Genus *Oceanodroma* Reichenbach
    O. hubbsi L. Miller 1951
Family Pelecanoididae (Gray). **Diving Petrels.**
Order Pelecaniformes Sharpe.
  Family Phaethontidae. **Tropicbirds.**
    Genus *Heliadornis* Olson
      H. ashbyi Olson 1985
  Family Plotopteridae Howard.
    Genus *Plotopterum* Howard
      P. joaquinensis Howard 1969
  Family Phalacrocoracidae Bonaparte. **Cormorants.**
    Genus *Phalacrocorax* Brisson
      P. auritus (Lesson 1831)
      P. femoralis L. Miller 1929
      P. goletensis Howard 1965
      P. idahensis (Marsh 1870)
      P. kennelli Howard 1949
      P. leptopus Brodkorb 1961

    P. macer Brodkorb 1958
    P. wetmorei Brodkorb 1955
Family Anhingidae Ridgway. Anhingas.
  Genus Anhinga Brisson
    A. grandis Martin and Mengel 1975
    A. subvolans (Brodkorb 1956)
Family Sulidae (Reichenbach). Boobies and Gannets.
  Genus Sula Brisson
    S. guano Brodkorb 1955
    S. humeralis L. Miller and Bowman 1958
    S. phosphata Brodkorb 1955
    S. pohli Howard 1958
    S. universitatis Brodkorb 1963
    S. willetti L. Miller 1925
  Genus Microsula Wetmore
    M. avita (Wetmore 1938)
  Genus Morus Vieillot
    M. lompocanus (L. Miller 1925)
    M. loxostyla (Cope 1870)
    M. magnus Howard 1978
    M. peninsularis Brodkorb 1955
    M. vagabundus Wetmore 1930
  Genus Palaeosula Howard
    P. stocktoni (L. Miller 1935)
  Genus Miosula L. Miller
    M. media Miller 1925
    M. recentior Howard 1949
Family Pelagornithidae Fürbringer. Bony-toothed Birds.
  Genus Pseudodontornis Lambrecht
    Pseudodontornis sp.
  Genus Osteodontornis Howard 1957
    O. orri Howard 1957
  Genus Cyphornis Cope
    C. magnus Cope 1894

Family Pelecanidae Vigors. Pelicans.
   Genus Pelecanus Linnaeus
     P. erythrorhynchos Gmelin 1789
     P. halieus Wetmore 1933
Order Ardeiformes (Wagler).
  Family Plataleidae Bonaparte. Ibises and Spoonbills.
   Genus Eudocimus Wagler
    Eudocimus sp.
   Genus Plegadis Kaup
     P. pharangites Olson 1981
   Genus Phimosus Wagler
    Phimosus sp.
   Genus Mesembrinibis Peters
    Mesembrinibis sp.
  Family Ardeidae Vigors. Herons.
   Genus Ardea Linnaeus
     A. alba Linnaeus 1758
     A. polkensis Brodkorb 1955
   Genus Egretta Forster
     E. subfluvia Becker 1985
   Genus Ardeola Boie
     A. validipes (Campbell 1976)
   Genus Nycticorax Forster
     N. fidens Brodkorb 1963
   Genus Botaurus Stephens
     B. hibbardi Moseley and Feduccia 1975
  Family Ciconiidae (Gray). Storks.
   Genus Propelargus Lydekker
     "P". olseni Brodkorb 1963
   Genus Mycteria Linnaeus
    Mycteria sp.
   Genus Ciconia Brisson
     C. maltha L. Miller 1910
   Genus Dissourodes Short
     D. milleri Short 1966

Order Anseriformes (Wagler).
  Family Anatidae Vigors. Ducks, Geese, and Swans
    Genus <u>Cygnus</u> Bechstein
      <u>Cygnus</u> sp.
    Genus <u>Olor</u> Wagler
      <u>O</u>. <u>hibbardi</u> (Brodkorb 1958)
    Genus <u>Paracygnus</u> Short
      <u>P</u>. <u>plattensis</u> Short 1969
    Genus <u>Anser</u> Brisson
      <u>A</u>. <u>pressus</u> (Wetmore 1933)
      <u>A</u>. <u>thompsoni</u> Martin and Mengel 1980
    Genus <u>Presbychen</u> Wetmore
      <u>P</u>. <u>abavus</u> Wetmore 1930
    Genus <u>Heterochen</u> Short
      <u>H</u>. <u>pratensis</u> Short 1970
    Genus <u>Eremochen</u> Brodkorb
      <u>E</u>. <u>russelli</u> Brodkorb 1961
    Genus <u>Branta</u> Scopoli
      <u>B</u>. <u>esmeralda</u> Burt 1929
      <u>B</u>. <u>howardae</u> L. Miller 1930
    Genus <u>Dendrochen</u> A. H. Miller
      <u>D</u>. <u>robusta</u> A. Miller 1944
    Genus <u>Dendrocygna</u> Swainson
      <u>D</u>. <u>eversa</u> Wetmore 1924
    Genus <u>Anas</u> Linnaeus
      <u>A</u>. <u>bunkeri</u> (Wetmore 1944)
      <u>A</u>. <u>crecca</u> Linnaeus 1758
      <u>A</u>. <u>greeni</u> (Brodkorb 1964)
      <u>A</u>. <u>integra</u> A. Miller 1944
      <u>A</u>. <u>ogallalae</u> (Brodkorb 1962)
      <u>A</u>. <u>pullulans</u> Brodkorb 1961
      <u>A</u>. <u>platyrhynchos</u> Linnaeus 1758
      <u>A</u>. <u>acuta</u> Linnaeus 1758
      <u>A</u>. <u>clypeata</u> Linnaeus 1758

Genus **Wasonaka** Howard
    W. yepomerae Howard 1966
Genus *Anabernicula* Ross
    A. minuscula (Wetmore 1924)
Genus *Brantadorna* Howard
    B. downsi Howard 1963
Genus *Aythya* Boie
    A. affinus (Eyton 1838)
Genus *Paranyroca* A. H. Miller and Compton
    P. magna A. H. Miller and Compton 1939
Genus *Ocyplonessa* Brodkorb
    O. shotwelli Brodkorb 1961
Genus *Melanitta* Boie
    M. perspicillata (Linnaeus 1758)
Genus *Bucephala* Baird
    B. albeola (Linnaeus 1758)
    B. fossilis Howard 1963
    B. ossivallis Brodkorb 1955
Genus *Mergus* Linnaeus
    M. miscellus Alvarez and Olson 1978
    M. merganser Linnaeus 1758
Genus *Oxyura* Bonaparte
    O. dominica Linnaeus 1766
    O. bessomi Howard 1963
Order Acciptriformes (Vieillot).
  Family Teratornithidae L. Miller. Teratorns.
    Genus *Teratornis* L. Miller
      T. incredibilis Howard 1952
  Family Vulturidae (Illiger). New World Vultures.
    Genus *Sarcoramphus* Duméril
      S. kernensis (L. Miller 1931)
    Genus *Pliogyps* Tordoff
      P. fisheri Tordoff 1959
      P. charon Becker 1986

Family Pandionidae (Sclater and Salvin). Ospreys.
  Genus <u>Pandion</u> Linnaeus
    <u>P</u>. <u>homalopteron</u> Warter 1976
    <u>P</u>. <u>lovensis</u> Becker 1985
Family Accipitridae (Vieillot). Hawks and Buzzards.
  Genus <u>Buteo</u> Lacêpède
    <u>B</u>. <u>ales</u> Wetmore 1926
    <u>B</u>. <u>conterminus</u> Wetmore 1923
    <u>B</u>. <u>contortus</u> Wetmore 1923
    <u>B</u>. <u>dananus</u> (Marsh 1871)
    <u>B</u>. <u>jamaicensis</u> (Gmelin 1788)
    <u>B</u>. <u>typhoius</u> Wetmore 1923
  Genus <u>Palaeastur</u> Wetmore
    <u>P</u>. <u>atavus</u> Wetmore 1923
  Genus <u>Miohierax</u> Howard
    <u>M</u>. <u>stocki</u> Howard 1944
  Genus <u>Parabuteo</u> Ridgway
    <u>Parabuteo</u> sp.
  Genus <u>Haliaeetus</u> Savigny
    <u>Haliaeetus</u> sp.
  Genus <u>Aquila</u> Brisson
    <u>A</u>. <u>chrysaetos</u> (Linnaeus 1758)
  Genus <u>Spizaetus</u> Vieillot
    <u>S</u>. <u>schultzi</u> Martin 1975
  Genus <u>Buteogallus</u> Lesson
    <u>B</u>. <u>enecta</u> (Wetmore 1923)
  Genus <u>Promilio</u> Wetmore
    <u>P</u>. <u>brodkorbi</u> Wetmore 1958
    <u>P</u>. <u>epileus</u> Wetmore 1958
    <u>P</u>. <u>floridanus</u> (Brodkorb 1956)
  Genus <u>Proictinia</u> Shufeldt
    <u>P</u>. <u>effera</u> Wetmore 1923
    <u>P</u>. <u>gilmorei</u> Shufeldt 1913

Genus *Palaeoborus* Coues
  P. rosatus A. H. Miller & Compton 1939
  P. umbrosus (Cope 1874)
Genus *Neophrontops* L. Miller
  N. dakotensis Compton 1935
  N. ricardoensis Rich 1980
  N. slaughteri Feduccia 1974
  N. vetustus Wetmore 1943
  N. vallecitoensis Howard 1963
Genus *Circus* Lacépède
  Circus sp.
Genus *Accipiter* Brisson
  Accipiter sp.
Family Falconidae Vigors.   Falcons.
  Genus *Falco* Linnaeus
    (?) *Falco* sp.
  Genus *Pediohierax* Becker
    P. ramenta (Wetmore 1936)
rder Galliformes (Temminck).
  Family Cracidae Vigors.   Chachalacas and Guans.
    Genus *Boreortalis* Brodkorb
      B. laesslei Brodkorb 1954
      B. phengites (Wetmore 1923)
      B. pollicaris A. H. Miller 1944
      B. tantala (Wetmore 1933)
      B. tedfordi (L. Miller 1952)
    Genus *Ortalis* Merrem
      O. affinis Feduccia & R. Wilson 1967
  Family Phasianidae Vigors.   Pheasants, Quail and Turkeys.
    Genus *Miortyx* A. H. Miller
      M. teres A. H. Miller 1944
    Genus *Cyrtonyx* Gould
      C. cooki Wetmore 1934

Genus _Lophortyx_ Bonaparte
- _L_. _shotwelli_ Brodkorb 1958
- _L_. _gambeli_ Gambel 1843

Genus _Colinus_ Goldfuss
- _C_. _hibbardi_ Wetmore 1944
- _C_. _suilium_ Brodkorb 1959

Genus _Palaealectoris_ Wetmore
- _P_. _incertus_ Wetmore 1930

Genus _Tympanuchus_ Gloger
- _T_. _stirtoni_ A. Miller 1944

Genus _Archaeophasianus_ Lambrecht
- _A_. _mioceanus_ (Shufeldt 1915)
- _A_. _roberti_ (Stone 1915)

Genus _Rhegminornis_ Wetmore
- _R_. _calobates_ Wetmore 1943

Genus _Proagriocharis_ Martin and Tate
- _P_. _kimballensis_ Martin and Tate 1970

Genus _Meleagris_ Linnaeus
- _M_. _gallopavo_ Linnaeus 1758
- _M_. _leopoldi_ A. H. Miller and Bowman 1956
- _M_. _progenes_ (Brodkorb 1964)
- _M_. _anza_ Howard 1963

Order Ralliformes (Reichenbach).
Family Rallidae Vigors. Rails, Gallinules, and Coots.

Genus _Rallus_ Linnaeus
- _R_. _elegans_ Audubon 1834
- _R_. _longirostris_ Boddaert 1783
- _R_. _lacustris_ (Brodkorb 1958)
- _R_. _phillipsi_ Wetmore 1957
- _R_. _prenticei_ Wetmore 1944
- _R_. _limicola_ Vieillot 1819

Genus _Creccoides_ Shufeldt
- _C_. _osbornii_ Shufeldt 1892

Genus _Coturnicops_ G. R. Gray
- _C_. _avita_ Feduccia 1968

    Genus *Gallinula* Brisson
        G. kansarum Brodkorb 1967
    Genus *Laterallus* G. R. Gray
        L. insignis Feduccia 1968
    Genus *Fulica* Linnaeus
        F. infelix Brodkorb 1961
        F. americana Gmelin 1789
        F. hesterna Howard 1963
  Family Gruidae Vigors. Cranes.
    Genus *Probalearica* Lambrecht
        P. crataegensis Brodkorb 1963
    Genus *Aramornis* Wetmore
        A. longurio Wetmore 1926
    Genus *Balearica* Brisson
        Balearica sp.
    Genus *Grus* Pallas
        G. americana (Linnaeus 1758)
        G. conferta A. Miller & Sibley 1942
        G. nannodes Wetmore & Martin 1930
        G. canadensis (Linnaeus 1758)
        G. haydeni Marsh 1870
  Family Phorusrhacidae (Ameghino)
    Genus *Titanis* Brodkorb
        T. walleri Brodkorb 1963
Order Charadriiformes (Huxley).
  Family Charadriidae Vigors. Plovers.
    Genus *Pluvialis* Brisson
        P. squatarola (Linnaeus 1758)
    Genus *Charadrius* Linnaeus
        C. vociferus Linnaeus 1758
  Family Scolopacidae Vigors. Sandpipers.
    Genus *Limosa* Brisson
        L. ossivallis Brodkorb 1967
        L. vanrossemi L. Miller 1925

Genus *Bartramia* Lesson
   B. umatilla Brodkorb 1958
Genus *Tringa* Linnaeus
   T. antiqua Feduccia 1970
Genus *Actitis* Illiger
   Actitis sp.
Genus *Arenaria* Brisson
   Arenaria sp.
Genus *Calidris* Merrem
   C. pacis Brodkorb 1955
   C. penepusilla (Brodkorb 1955)
   C. rayi (Brodkorb 1963)
Genus *Micropalama* Baird
   M. hesternus Wetmore 1924
Genus *Philomachus* Merrem
   Philomachus sp.
Family Phoenicopteridae Bonaparte. Flamingos.
   Genus *Phoenicopterus* Linnaeus
     P. floridanus Brodkorb 1953
     P. stocki Miller 1944
   Genus *Megapaloelodus* A. H. Miller
     M. connectens A. Miller 1944
     M. opsigonus Brodkorb 1961
Family Haematopodidae Gray. Oystercatchers.
   Genus *Haematopus* Linnaeus
     H. sulcatus (Brodkorb 1955)
Family Recurvirostridae Bonaparte. Avocets and Stilts.
   Genus *Himantopus* Brisson
     Himantopus sp.
   Genus *Recurvirostra* Linnaeus
     Recurvirostra sp.
Family Jacanidae Stejneger. Lily-trotters.
   Genus *Jacana* Brisson
     J. farrandi Olson 1976

Family Burhinidae Mathews. Thick-knees.
  Genus Burhinus Illiger
    B. lucorum Bickart 1981
Family Glareolidae Brehm. Pratincoles and Coursers.
  Genus Paractiornis Wetmore
    P. perpusillus (Wetmore 1930)
Family Laridae Vigors. Gulls and Terns.
  Genus Larus Linnaeus
    L. elmorei Brodkorb 1955
  Genus Gaviota A. H. Miller and Sibley
    G. niobrara A. Miller and Sibley 1941
  Genus Rissa Stephens
    Rissa sp.
  Genus Sterna Linnaeus
    Sterna sp.
Family Stercorariidae (Gray). Jaegers and Skuas.
  Genus Catharacta Brünnich
    Catharacta sp.
  Genus Stercorarius Brisson
    Stercorarius sp.
Family Alcidae Vigors. Auks.
  Genus Alle Link
    Alle sp.
  Genus Pinguinus Bonnaterre
    P. alfrednewtoni Olson 1977
  Genus Alca Linnaeus
    Alca sp.
  Genus Miocepphus Wetmore
    M. mcclungi Wetmore 1940
  Genus Cepphus Pallas
    C. olsoni Howard 1982
  Genus Cerorhinca Bonaparte
    C. dubia L. Miller 1925
    C. minor Howard 1971

Genus **Brachyramphus** M. Brandt
  B. pliocenus Howard 1949
Genus **Synthlioboramphus** M. Brandt
  Synthlioboramphus sp.
Genus **Ptychoramphus** M. Brandt
  P. tenius L. Miller and Bowman 1958
Genus **Cyclorrhynchus** Kaup
  Cyclorrhynchus sp.
Genus **Uria** Brisson
  U. brodkorbi Howard 1981
  U. paleohesperis Howard 1982
Genus **Australca** Brodkorb
  A. antiqua (Marsh 1870)
  A. grandis Brodkorb 1955
Genus **Aethia** Merrem
  A. rossmoori Howard 1968
Genus **Fratercula** Brisson
  Fratercula sp.
Genus **Praemancalla** Howard
  P. lagunensis Howard 1966
  P. wetmorei Howard 1976
Genus **Alcodes** Howard
  A. ulnulus Howard 1968
Genus **Mancalla** Lucas
  M. californiensis Lucas 1901
  M. cedrosensis Howard 1971
  M. diegensis (L. Miller 1937)
  M. emlongi Olson 1981
  M. milleri Howard 1970
Order Columbiformes (Latham).
  Family Columbidae (Illiger). Doves.
    Genus **Zenaida** Bonaparte
      Z. prior (Brodkorb 1969)

Order Cuculiformes (Wagler).
  Family Cuculidae Vigors. Cuckoos and Roadrunners.
    Genus *Cursoricoccyx* Martin & Mengel
      C. *geraldinae* Martin & Mengel, 1984
Order Psittaciformes (Wagler).
  Family Psittacidae (Illiger). Parrots.
    Genus *Conuropsis* Salvadori
      C. *fratercula* Wetmore 1926
Order Strigiformes (Wagler).
  Family Strigidae Vigors. Owls.
    Genus *Strix* Linnaeus
      S. *dakota* A. Miller 1944
    Genus *Asio* Brisson
      A. *brevipes* Ford and Murray 1967
    Genus *Bubo* Duméril
      Bubo sp.
    Genus *Otus* Pennant
      Otus sp. near *O*. *asio* (Linnaeus 1758)
      Otus sp. near *O*. *flammeolus* (Kaup 1853)
    Genus *Speotyto* Gloger
      S. *megalopeza* Ford 1966
  Family Tytonidae Ridgway. Barn Owls.
Order Coraciiformes Forbes.
  Family Momotidae (Gray). Motmots.
Order Piciformes (Meyer and Wolf).
  Family Capitonidae Bonaparte. Barbets.
  Family Picidae Vigors. Woodpeckers
    Genus *Pliopicus* Feduccia and Wilson
      P. *brodkorbi* Feduccia & R. Wilson 1967
    Genus *Palaeonerpes* Cracraft and Morony
      P. *shorti* Cracraft & Morony 1969
    Genus *Campephilus* G. R. Gray
      C. *dalquesti* Brodkorb 1971
    Genus *Colaptes* Vigors
      Colaptes sp.

Order Passeriformes (Linnaeus).
   Family Corvidae Vigors. Crows and Jays.
      Genus *Miocitta* Brodkorb
         M. galbreathi Brodkorb 1972
      Genus *Corvus* Linnaeus
         Corvus sp.
      Genus *Protocitta* Brodkorb
         P. ajax Brodkorb 1972
   Family Palaeoscinidae Howard.
      Genus *Palaeoscinis* Howard
         P. turdirostris Howard 1957
   Family Hirundinidae Vigors. Swallows.
      Genus *Hirundo* Linnaeus
         H. aprica Feduccia 1967
   Family Parulidae American Ornithologists' Union. Wood Warblers.
   Family Emberizidae Vigors. Sparrows.
      Genus *Ammodramus* Swainson
         A. hatcheri (Shufeldt 1913)
      Genus *Passerina* Vieillot
         Passerina sp.
      Genus *Junco* Wagler
         Junco sp.
   Family Fringillidae

North American Neogene Avian Localities

Late Arikareean

Agate Fossil Quarries
   Formation: Upper Harrison
   State: Nebraska, Sioux Co.
   NALMA or Correlative: Late Arikareean
   Radiometric Dates: 21.3 mybp on the Agate Ash in the upper part of the Harrison Fm., and therefore a maximum date.
   General References: Hunt 1972, 1981
   Avian References: Wetmore 1923, 1926b, 1928b, 1930a, 1933a, 1943c, 1958; Olson and Steadman 1978
   Avian Taxa: Accipitridae
            *_Buteo_ _ales_
            Buteo _typhoius_
            Buteo sp.
            *_Palaeastur_ _atavus_
            Proictinia _effera_
            Accipitrid, indet. (large)
         Cracidae
            *_Boreortalis_ _tantala_
         Phasianidae
            *_Palaealectoris_ _incertus_
         Glareolidae
            *_Paractiornis_ _perpusillus_
   Comments: Hunt (1981) showed that most of the Agate Spring Quarries are in the base of the Upper Harrison Beds (= Marsland Fm.), not in the Harrison Formation as previously had been believed.

Pyramid Hill
- Formation: Jewett Sand, Pyramid Hill Sand Member
- State: California, Kern Co.
- NALMA or Correlative: Late Arikareean ("Vaqueros")
- Radiometric Dates: none
- General References: Kellogg 1932, Wilson 1935, Savage and Barnes 1972, Mitchell and Tedford 1973, Barnes 1979
- Avian References: Howard 1969, 1972
- Avian Taxa: Plotopteridae
  - *Plotopterum joaquinensis*

Vasquez Canyon
- Formation: Tick Canyon
- State: California, Los Angeles Co.
- NALMA or Correlative: Late Arikareean
- Radiometric Dates: none
- General References: Jahns 1940
- Avian References: Howard 1944
- Avian Taxa: Accipitridae
  - *Miohierax stocki*

Hemingfordian

Astoria
    Formation: Astoria
    State: Oregon, Lincoln Co.
    NALMA or Correlative: Pillarian, upper part
    Radiometric Dates: none
    General References: Armentrout 1981
    Avian References: Olson 1985b
    Avian Taxa: Pelagornithidae
                  <u>Pseudodontornis</u> sp.
              Diomedeidae
                  <u>Diomedea</u> sp.

Carmanah Point
    Formation: Carmanah Point Beds
    State: Vancouver Is., British Columbia, Canada
    NALMA or Correlative: (?) Late Hemingfordian or Barstovian
    Radiometric Dates: none
    General References: Merriam 1896, Russell 1968
    Avian References: Cope 1894, Wetmore 1928c
    Avian Taxa: Pelagornithidae
                  *<u>Cyphornis</u> <u>magnus</u>
    Comments: The stratigraphic provenance of this species is uncertain. Wetmore (1928c) followed Merriam (1896) in considering the Carmanah Point Beds to be most similar to the Miocene Astoria Fm., rather than to the Oligocene Sooke Formation. If this is the case, then this locality is "Temblor" in age, and would be roughly equivalent to the late Hemingfordian or Barstovian. Based on the marine mammals from the Olympic Peninsula, Ray (pers. comm.) feels that it is more likely that the "Carmanah Point Beds" are correlative to the Sooke

Formation. See Domning et al. (1986) and references therein for a discussion of geology of the northern part of the Olympic peninsula and Armentrout (1981) for the correlation and ages of Cenozoic chronostratigraphic units in the northern Pacific.

Dunlap Camel Quarry
    Formation: Runningwater
    State: Nebraska, Dawes Co.
    NALMA or Correlative: Early Hemingfordian
    Radiometric Dates: none
    General References: Schultz and Falkenbach 1940
    Avian References: none
    Avian Taxa: Gruidae
              Balearicinae
    Comments: This record is based on unstudied material in the Frick Collections.

Farmingdale
    Formation: Kirkwood, (Ammodon Beds)
    State: New Jersey, Monmouth Co.
    NALMA or Correlative: Early Hemingfordian
    Radiometric Dates: none
    General References: Marsh 1893, Tedford and Hunter 1984
    Avian References: Shufeldt 1915, Wetmore 1926c
    Avian Taxa: Sulidae
              Morus loxostyla
    Comments: Shufeldt (1915) described *Sula atlantica*, later placed in synonymy with *Morus loxostyla* by Wetmore (1926c), from the Marsh collections of Yale University. These early collections came from the basal sands of the Asbury Park Member of the Kirkwood Formation in the vicinity of Farmingdale

(Tedford and Hunter 1984:134). Because the type of *Morus loxostyla* from the Calvert Formation has been lost for the last 75 years, all comparisons between it and "*Sula atlantica*" have been made only with Cope's (1870) original description and illustrations.

Flint Hill
    Formation: Batesland
    State: South Dakota, Bennett Co.
    NALMA or Correlative: Early Hemingfordian
    Radiometric Dates: none
    General References: Harksen and Macdonald 1967; J. Martin 1985
    Avian References: A. Miller and Compton 1939, A. Miller 1944
    Avian Taxa: Anatidae
        *<u>Paranyroca magna</u>
        *<u>Dendrochen robusta</u>
        *<u>Anas integra</u>
      Accipitridae
        *<u>Palaeoborus rosatus</u>
        Buteonine indet.
      Cracidae
        *<u>Boreortalis pollicaris</u>
      Phasianidae
        *<u>Tympanuchus stirtoni</u>
        *<u>Miortyx teres</u>
      Phoenicopteridae
        *<u>Megapaloelodus connectens</u>
      Strigidae
        *<u>Strix dakota</u>
    Comments: This locality is also known as Flint Hills North. J. Martin (1985) discusses the history of this locality and shows that it is correlative with

Martin Canyon Quarry A based on the small mammal fauna, although Flint Hills may be slightly younger. Ford (1967) questioned the generic identity of *Strix* *dakota*.

Foley Quarry
    Formation: Box Butte, Red Valley Mbr.
    State: Nebraska, Box Butte Co.
    NALMA or Correlative: Late Hemingfordian
    Radiometric Dates: none
    General References: Galusha 1975
    Avian References: Galusha 1975
    Avian Taxa: Accipitridae
    Comments: This record is based on unstudied, fragmentary material in the Frick Collections.

Ginn Quarry
    Formation: unnamed equivalent to the Sheep Creek Fm. of Sioux County.
    State: Nebraska, Dawes Co.
    NALMA or Correlative: Late Hemingfordian
    Radiometric Dates: none
    General References: Galusha 1975
    Avian References: none
    Avian Taxa: Accipitridae
                  Gruidae
                  Balearicinae
    Comments: These records are based on unstudied material in the Frick Collections. See Skinner et al. (1977) for a discussion of the localities in the Sheep Creek Formation.

Hogtown Creek and University of Florida Campus
    Formation: Hawthorn
    State: Florida, Alachua Co.

NALMA or Correlative: (?) Hemingfordian, see comments below.
Radiometric Dates: none
General References: Williams et al. 1977
Avian References: Brodkorb 1963b
Avian Taxa: Procellariidae
  *Puffinus micraulax
  Sulidae
  *Sula universitatis
Comments: Creeks in Gainesville have produced a mixed faunal assemblage of Hemingfordian and Hemphillian age land mammals. Because the avian holotypes were found in an active streambed, there is no direct stratigraphic evidence that they are from the Hemingfordian age Hawthorn Formation (although this is likely). Sula universitatis is known from only a incomplete, waterworn carpometacarpus and its systematic position needs verification.

Martin Canyon Quarry A
Formation: Martin Canyon Beds
State: Colorado, Logan Co.
NALMA or Correlative: Early Hemingfordian
Radiometric Dates: none
General References: Wilson 1960, J. Martin 1978, L. Martin and Mengel 1984
Avian References: L. Martin and Mengel 1984
Avian Taxa: Cracidae
  Genus and species indet.
  Cuculidae
  *Cursoricoccyx geraldinae
Comments: This local fauna is correlative to the Flint Hills local fauna in the Batesland Fm. (see above) and includes additional unstudied species of birds (L. Martin and Mengel 1984).

Nye
    Formation: Nye
    State: Oregon, Lincoln Co.
    NALMA or Correlative: Newportian, lower part
        (equivalent to late Hemingfordian and early
        Barstovian)
    Radiometric Dates: none
    General References: Armentrout 1981
    Avian References: Olson 1985b
    Avian Taxa: Pelagornithidae
                    Osteodontornis sp.

Red Valley Fill
    Formation: Box Butte
    State: Nebraska, Box Butte Co.
    NALMA or Correlative: Late Hemingfordian
    Radiometric Dates: none
    General References: Galusha 1975
    Avian References: Galusha 1975
    Avian Taxa: Strigiformes
    Comments: This record is based on unstudied material
        in the Frick Collections.

Seaboard
    Formation: Torreya
    State: Florida, Leon Co.
    NALMA or Correlative: Mid Hemingfordian
    Radiometric Dates: none
    General References: Olsen 1964, Hunter and Huddleston
        1982, Tedford and Hunter 1984
    Avian References: Brodkorb 1963a
    Avian Taxa: Ciconiidae
                    *Propelargus olseni
                Gruidae
                    *Probalearica crataegensis

32

Comments: This local fauna occurs in a thin remnant of the Pododesmus scopelus Assemblage Zone in the lower part of the Torreya Formation and contains mammalian taxa apparently identical with those from the Thomas Farm local fauna (Tedford and Hunter 1984).

"Scottsbluff or Chimney Rock"
    Formation: Brule or lower Arikaree Group
    State: Nebraska, "Scotts Bluff or Morill Co."
    NALMA or Correlative: Orellan through early Arikareean
    Radiometric Dates: none
    General References: none
    Avian References: Shufeldt 1915, Lambrecht 1933, Brodkorb 1964b
    Avian Taxa: Phasianidae
                *Archaeophasianus mioceanus
    Comments: Shufeldt (1915) described Phasianus mioceanus from two cotypes in the Marsh collection of Yale University, giving the type locality as only "Chimney Rock and Scott's Bluff, Nebraska." This species and Phasianus roberti Stone from the John Day Formation were later moved to Archaeophasianus (Lambrecht 1933). Brodkorb (1964b) gave a locality description for Archaeophasianus mioceanus that included a county designation and reported the specimens as coming from the Sheep Creek or Marsland Formations. Brodkorb correctly (1964b) noted that these localities are about 20 miles apart (even though Shufeldt questionably considered the two specimens to be from the same individual). Tedford (pers. comm.) considers the Scottsbluff and Chimney Rock localities to be from either the Brule Formation or

the lower Arikaree Group, thus the actual geologic age of this "species", depending on which specimen is designated as a holotype, range from the Orellan through the early Arikareean.

Sheep Creek
    Formation: Sheep Creek
    State: Nebraska, Sioux Co.
    NALMA or Correlative: Late Hemingfordian
    Radiometric Dates: 16.1 ± 3.7 mybp
    General References: Matthew 1924, Cook and Cook 1933, Skinner et al. 1977
    Avian References: Wetmore 1923, 1926d, 1934, 1936, 1943c, Bickart 1981, Becker 1987a
    Avian Taxa: Anatidae
            Falconidae
                *Pediohierax ramenta
            Accipitridae
                *Neophrontops vetustus
                *Buteogallus enecta
                Genus and sp. to be determined
            Phasianidae
                *Cyrtonyx cooki
            Rallidae
            Gruidae
                *Aramornis longurio
            (?) Jacanidae
            Burhinidae
                *Burhinus lucorum
            Psittacidae
                *Conuropsis fratercula
            (?) Strigiformes
    Comments: In addition to published taxa, the above list includes unstudied material from Long, Greenside, Thomson (Stonehouse Draw), Thistle (Antelope

Draw), and Hilltop quarries in the Frick collections of the American Museum.

Thomas Farm
    Formation: Hawthorn
    State: Florida, Gilchrist Co.
    NALMA or Correlative: Mid Hemingfordian
    Radiometric Dates: none
    General References: Simpson 1932, White 1942, Patton 1967, Tedford and Frailey 1976
    Avian References: Wetmore 1943b, 1958, Brodkorb 1954, 1956, 1963b, Becker 1986b, Cracraft 1971, Olson and Farrand 1974, Steadman 1980
    Avian Taxa: Anhingidae
            *Anhinga subvolans
        Accipitridae
            *Promilio floridanus
            *P. epileus
            *P. brodkorbi
            (?) Buteo sp.
        Cracidae
            *Boreortalis laesslei
        Phasianidae
            *Rhegminornis calobates
        Columbidae, 2 sp.
        Coraciiformes
        Capitonidae
        Parulidae
    Comments: The Columbidae, Coraciiformes, Capitonidae and Parulidae are reported as preliminary identifications.

Barstovian

Barstow
- Formation: Barstow
- State: California, San Bernardino Co.
- NALMA or Correlative: Early to Late Barstovian
- Radiometric Dates: 16.3 ± 0.3 mybp near the base of the faunal sequence; 14.8 ± 0.3 and 15.5 mybp near the division of the Green Hills and Barstow faunas; and 13.4 ± 0.7 mybp near the top of the Barstow fauna.
- General References: Merriam 1919, Lindsay 1972, Woodburne et al. 1981, Tedford et al. in press
- Avian References: L. Miller 1950, 1952, 1966, Holman 1961
- Avian Taxa: Podicipedidae
  Anatidae
     5 species
  Accipitridae
     Buteonine, 2 sp.
  Cracidae
     *<u>Boreortalis</u> <u>tedfordi</u>
  Gruidae
  Phoenicopteridae
     <u>Megapaloelodus</u> <u>connectens</u>
  Lari
  Strigiformes
- Comments: "Barstow" is used here in the broad sense, and includes birds from both the Barstow and the Green Hills local faunas. The faunal list also includes some unstudied material in the Frick Collections from the Upper Steepside Quarry and Sunder Ridge.

Calabasas
    Formation: Modelo
    State: California, Los Angeles Co.
    NALMA or Correlative: Barstovian or Clarendonian equivalent
    Radiometric Dates: none
    General References: none
    Avian References: L. Miller 1929
    Avian Taxa: Phalacrocoracidae
                  *Phalacrocorax femoralis
    Comments: See Miller and DeMay (1942) for a description of this locality and other references to this species.

Calvert
    Formation: Calvert
    State: Virginia, Maryland
    NALMA or Correlative: Langhian (early Barstovian equivalent)
    Radiometric Dates: none
    General References: Gazin and Collins 1950, Blackwelder and Ward 1976, Tedford and Hunter 1984
    Avian References: Wetmore 1926c, 1930a, 1938, 1940, 1941, Alvarez and Olson 1978, Olson 1984, 1985a
    Avian Taxa: Gaviidae
                Gavia sp.
           Diomedeidae
           Procellariidae
              *Puffinus conradi
              Puffinus sp.
           Phaethontidae
              *Heliadornis ashbyi
           Sulidae
              *Morus loxostyla
              Morus sp.

*Microsula avita
Pelagornithidae (2 species)
Anatidae
　Tadornini
　*Mergus miscellus
Phasianidae (2 species)
　Palaealectoris sp.
Stercorariidae
　Genus indet., 2 sp.
Alcidae
　*Miocepphus mcclungi
　Miocepphus sp.
Charadriiformes
　Family indet.

Comments: Most of the specimens of fossil birds come from lithostratigraphic "Zones" 10 to 13 of Shattuck (1904). Mergus miscellus is from basal part of Zone 13 (Alvarez and Olson 1978); Heliadornis ashbyi is from Zone 11, about 60 cm below the base of Zone 12 (Olson 1985a). Fulmarus sp. and Columbidae were both reported in error from this local fauna. It appears that the composite assemblage of birds originate in a slightly lower stratigraphic position within the Calvert Formation than does the mammalian assemblage (Chesapeake Bay Fauna from "Zones" 13-15; Tedford and Hunter 1984). Olson (pers. comm.) notes that there seems to be little change in the composition of the avifauna throughout the sampled portions of this formation.

Crookston Bridge
　Formation: Valentine, Crookston Bridge Mbr.
　State: Nebraska, Cherry Co.
　NALMA or Correlative: Late Barstovian
　Radiometric Dates: none

General References: Skinner et al. 1968
Avian References: Short 1966, Skinner and Johnson 1984
Avian Taxa: Ciconiidae
      *Dissourodes milleri
      Anatidae
      Accipitridae
      Galliformes

Comments: This list includes unstudied material from the Frick Crookston Bridge, Nenzel, and Schoetiger quarries. The Crookston Bridge Member overlies the Cornell Dam Member (Skinner and Johnson 1984). Dissourodes milleri is from Railway Quarry "A", one of the major Valentine localities.

Devils Gulch
    Formation: Valentine, Devils Gulch Mbr.
    State: Nebraska, Brown County
    NALMA or Correlative: Late Barstovian
    Radiometric Dates: none
    General References: Barbour 1913, Woodburne et al. 1981, Skinner and Johnson 1984
    Avian References: Short 1970
    Avian Taxa: Anatidae
          *Heterochen pratensis
        Gruidae
          Balearicinae
        Picidae

Comments: This list includes material from the Frick Rattlesnake Gulch and Devils Gulch Quarries, the latter being the type locality of Heterochen pratensis.

Garner Bridge
    Formation: Valentine
    State: Nebraska, Cherry Co.
    NALMA or Correlative: Late Barstovian
    Radiometric Dates: none
    General References: none
    Avian References: none
    Avian Taxa: Accipitridae
               Galliformes
    Comments: These records are based on unstudied material in the Frick Collections.

Kennesaw
    Formation: Pawnee Creek
    State: Colorado, Logan Co.
    NALMA or Correlative: Late Barstovian
    Radiometric Dates: none
    General References: Galbreath 1953, 1964, Wilson 1960
    Avian References: Brodkorb 1972
    Avian Taxa: Corvidae
              *<u>Miocitta galbreathi</u>

Lower Snake Creek
    Formation: Olcott
    State: Nebraska, Sioux Co.
    NALMA or Correlative: Early Barstovian
    Radiometric Dates: none
    General References: Matthew 1924, Matthew & Cook 1909, Skinner et al. 1977 and references therein.
    Avian References: Wetmore 1923, 1926a, 1928a, Becker 1987a
    Avian Taxa: Anhingidae or Phalacrocoracidae
               Anatidae
                   Anserinae
                   Anatinae

        Accipitridae
           *<u>Buteo</u> <u>typhoius</u>
           *<u>Buteo</u> <u>contortus</u>
        Falconidae
           <u>Pediohierax</u> <u>ramenta</u>
        Galliformes
           (3 sizes)
        Gruidae
           Balearicinae
        (?) Jacanidae
        Piciformes or Passeriformes
Comments: This list includes mostly unstudied material from the following Frick quarries: Sinclair Draw, Jenkins, East Sand, Version, New Surface, East Surface, Mill, Echo, Humbug, and Boulder.

Norden Bridge
    Formation: Valentine, Cornell Dam Mbr.
    State: Nebraska, Brown and Keya Paha Co.
    NALMA or Correlative: Late Barstovian
    Radiometric Dates: 13.6 ± 1.2 mybp; (Hurlbut Ash; Skinner and Johnson 1984)
    General References: Skinner and Johnson 1984
    Avian References: none
    Avian Taxa: Anatidae
                  Anserinae
           Rallidae
    Comments: These records are based on unstudied material from the Frick Norden Bridge and Egelhof quarries.

Observation Quarry
    Formation: Sand Canyon Beds
    State: Nebraska, Dawes Co.

NALMA or Correlative: Early Barstovian
Radiometric Dates: 14.2 ± 1.4 mybp
General References: Elias 1942, Tedford et al. in press
Avian References: Becker 1986a, 1987a
Avian Taxa: Ardeidae
    Ardea sp.
  Anatidae
    Anserinae
    (?) Anatinae
  Accipitridae (3 species)
  Falconidae
    Pediohierax ramenta
  Galliformes (3 species)
  Gruidae
    (?) Balearicinae
  Charadriiformes
  Strigiformes
  Passeriformes (or Piciformes)
Comments: This list includes unstudied material in the Frick Collections.

Pawnee Quarry
  Formation: Pawnee Creek
  State: Colorado, Weld Co.
  NALMA or Correlative: Early Barstovian
  Radiometric Dates: none
  General References: Tedford et al. in press
  Avian References: none
  Avian Taxa: Charadriiformes
  Comments: This record is based on unstudied material in the Frick Collections.

Paulina Creek
   Formation: Mascall
   State: Oregon, Grant Co.
   NALMA or Correlative: Early Barstovian
   Radiometric Dates: 15.8 mybp, low in section
   General References: Downs 1956
   Avian References: Shufeldt 1915, Stone 1915
   Avian Taxa: Phasianidae
            *Archaeophasianus roberti
   Comments: See remarks under "Scottsbluff or Chimney Rock".

Sharktooth Hill
   Formation: Round Mountain Silt
   State: California, Kern Co.
   NALMA or Correlative: "Temblor" (Late Barstovian)
   Radiometric Dates: none
   General References: Mitchell 1965, 1966, Addicott 1970, Barnes 1972, 1976, 1978, Savage and Barnes 1972, Repenning and Tedford 1977
   Avian References: Wetmore 1930b, L. Miller 1961, 1962, Howard 1966c, Warter 1976, Howard 1984
   Avian Taxa: Diomedeidae
                  *Diomedea californica
                  *D. milleri
               Procellariidae
                  *Fulmarus miocaenus
                  *Puffinus inceptor
                  *P. priscus
                  *P. mitchelli
               Ciconiidae
               Pelagornithidae
                  Osteodontornis ?orri
               Sulidae
                  *Morus vagabundus

43

              Morus sp.
          Vulturidae
          Anatidae
             *_Presbychen abavus_
             Branta sp.
          Pandionidae
             *_Pandion homalopteron_
          Phoenicopteridae
             Megapaloelodus sp.
          Recurvirostridae
             Recurvirostra sp.
Comments: See L. Miller and DeMay (1942) for additional references.

Skull Ridge
    Formation: Tesuque, Skull Ridge Mbr.
    State: New Mexico, Santa Fe Co.
    NALMA or Correlative: Early Barstovian
    Radiometric Dates: 14.1 $\pm$1.1; 14.6 $\pm$1.2 mybp fisson track ages (Izett and Naeser 1981) considered minimum by Barghoorn (1981).
    General References: Galusha and Blick 1971
    Avian References: Cope 1874
    Avian Taxa: Accipitridae
                *_Palaeoborus umbrosus_
                Additional Accipitridae
    Comments: This list also includes unstudied material in the Frick White Operation Quarry.

Trinity River
    Formation: Fleming
    State: Texas, San Jacinto Co.
    NALMA or Correlative: Early Barstovian
    Radiometric Dates: none
    General References: Tedford et al. in press
    Avian References: none
    Avian Taxa: Gruidae
    Comments: This record is based on unstudied material in the Frick Collections.

Clarendonian

"Ash Hollow"
    Formation: Ogallala group, undifferentiated
    State: South Dakota, Bennett Co.
    NALMA or Correlative: Clarendonian
    Radiometric Dates: none
    General References: see comments in Skinner et al. 1977
    Avian References: Brodkorb 1964a
    Avian Taxa: Anatidae
              *<u>Anas</u> <u>greeni</u>
    Comments: The holotype of this species is from SDSM Loc. V631, reported as being from the lower part of the Ash Hollow Formation. If this is correct, then the age is probably early Clarendonian. Although most localities in this part of South Dakota have been considered as coming from the Ash Hollow Formation, detailed stratigraphic studies of the sediments are lacking. Until such studies appear, it is best to consider these localities as being from undifferentiated sediments in the Valentine-Ash Hollow formations.

Big Spring Canyon
    Formation: Ogallala Group
    State: South Dakota, Bennett Co.
    NALMA or Correlative: Early Clarendonian
    Radiometric Dates: none
    General References: Gregory 1942, Skinner and Johnson 1984
    Avian References: Compton 1935a
    Avian Taxa: Anatidae
              <u>Branta</u> sp.
            Accipitridae

Neophrontops dakotensis
Comments: See remarks in Skinner and Johnson (1984) and above.

Black Butte
  Formation: Juntura (upper part)
  State: Oregon, Malheur Co.
  NALMA or Correlative: Mid-late Clarendonian
  Radiometric Dates: 12.4 mybp from basalt at the top of the lower member of the Juntura Formation; 9.4 $\pm$ 0.6 mybp from the lowest unit of the Drewsey Formation.
  General References: Shotwell 1963
  Avian References: Brodkorb 1961
  Avian Taxa: Phalacrocoracidae
              Phalacrocorax leptopus
            Anatidae
              *Anas pullulans
              *Eremochen russelli
              *Ocyplonessa shotwelli
            Rallidae
              *Fulica infelix
            Phoenicopteridae
              *Megapaloelodus opsigonus

Black Hawk Ranch
  Formation: Green Valley
  State: California, Contra Costa Co.
  NALMA or Correlative: Late Clarendonian (Montediablan, type fauna)
  Radiometric Dates: none
  General References: Macdonald 1948, Ritchey 1948, Webb and Woodburne 1964
  Avian References: A. Miller and Sibley 1942

Avian Taxa: Gruidae
            Grus conferta

Burge
    Formation: Valentine, Burge Mbr.
    State: Nebraska, Brown Co.
    NALMA or Correlative: Early Clarendonian
    Radiometric Dates: none
    General References: Skinner et al. 1968, Webb 1969a, Woodburne et al. 1981, Skinner and Johnson 1984
    Avian References: Webb 1969a
    Avian Taxa: Corvidae
                 Galliformes
    Comments: This list includes unstudied material from the Frick Lucht Quarry. Corvidae reported by Webb (1969a) from "Crazy Locality". Tedford et al. (in press) consider this local fauna to be latest Barstovian in age.

Cañana Pilares
    Formation: Zia Sand, unnamed upper member (sensu Tedford 1982)
    State: New Mexico, Sandoval Co.
    NALMA or Correlative: Late Barstovian (but see comments below)
    Radiometric Dates: none
    General References: Galusha 1966, Tedford 1981, 1982
    Avian References: none
    Avian Taxa: Anatidae
                 Phoenicopteridae
    Comments: These records are based on unstudied material in the Frick Collections. The specimens come from the lower green zone ("B") and from the upper algal limestones ("A") within the upper (unnamed) member of the Zia Sand (sensu Tedford 1982).

Clarendon
    Formation: Clarendon
    State: Texas, Donley Co.
    NALMA or Correlative: Clarendonian
    Radiometric Dates: none
    General References: see Webb (1969a) for list
    Avian References: none
    Avian Taxa: Accipitridae
              Phasianidae
                  Meleagridinae
              Phoenicopteridae
                  Megapaloelodus sp.
    Comments: This list includes unstudied material in the Frick collections.

Del Gado Drive (Sherman Oaks)
    Formation: Modelo, Upper Mbr.
    State: California, Los Angeles
    NALMA or Correlative: Clarendonian ("Margaritan")
    Radiometric Dates: none
    General References: Addicott 1972, Barnes 1976, Repenning and Tedford 1977
    Avian References: Howard 1962, Howard and White 1962
    Avian Taxa: Procellariidae
                  Puffinus diatomicus
              Pelagornithidae
                  Osteodontornis orri
              Sulidae
                  Sula willetti

Dove Spring
    Formation: Ricardo
    State: California, San Bernardino Co.
    NALMA or Correlative: Late Clarendonian (Montediablan)

Radiometric Dates: 10.0 mybp below local fauna
General References: Whistler 1970
Avian References: Rich 1980
Avian Taxa: Accipitridae
    Neophrontops ricardoensis
Comments: The highest Ricardo assemblage equates with faunas of Upper Ash Hollow Formation of Nebraska. See Ricardo (restricted) below.

Driftwood Creek
    Formation: "Beds equivalent to Ogallala Group"
    State: Nebraska, Hitchcock Co.
    NALMA or Correlative: late Barstovian or Clarendonian
    Radiometric Dates: none
    General References: none
    Avian References: Cracraft and Morony 1969
    Avian Taxa: Picidae
        Palaeonerpes shorti
    Comments: The stratigraphic position of the type locality within the Ogallala Group is unknown.

El Sereno
    Formation: Modelo
    State: California, Los Angeles Co.
    NALMA or Correlative: Clarendonian ("Margaritan")
    Radiometric Dates: none
    General References: Kellogg 1934, Barnes 1976
    Avian References: Howard 1958, Howard and White 1962
    Avian Taxa: Sulidae
        Palaeosula stocktoni

Fish Lake Valley
    Formation: Esmeralda
    State: Nevada, Esmeralda Co.
    NALMA OR Correlative: Early Clarendonian

Radiometric Dates: 11.8 mybp from middle of fossiliferous section; 11.4 and 11.7 mybp from just below micromammals.
General References: Mawby 1965, 1968a, 1968b, Suthard 1966, Tedford et al. in press
Avian References: Burt 1929
Avian Taxa: Anatidae
        *<u>Branta esmeralda</u>

Hollow Horn Bear Quarry
    Formation: Ogallala Group, undifferentiated
    State: South Dakota, Todd Co.
    NALMA or Correlative: Late Clarendonian
    Radiometric Dates: none
    General References: Skinner and Johnson 1984: 319
    Avian References: none
    Avian Taxa: Anatidae
                Anserinae
                Anatinae: Tadornini
            Galliformes
            Rallidae
            Gruidae
                Balearicinae
    Comments: This list is based on unstudied material in the Frick Collections.

Laguna Niguel
    Formation: Monterey
    State: California, Orange Co.
    NALMA or Correlative: Clarendonian
    Radiometric Dates: none
    General References: Savage and Barnes 1972, Barnes et al. in prep.
    Avian References: Howard 1976, 1978

Avian Taxa: Gaviidae
        *<u>Gavia brodkorbi</u>
    Diomedeidae
        <u>Diomedea</u> ?<u>californica</u>
        <u>Diomedea</u> sp.
    Procellariidae
        *<u>Puffinus barnesi</u>
    Oceanitidae
        <u>Oceanodroma</u> sp.
    Pelagornithidae
        <u>Osteodontornis orri</u>
    Sulidae
        <u>Morus lompocanus</u>
        *<u>Morus magnus</u>
        (?) <u>Miosula media</u>
        Sulidae spp. indet.
    Alcidae
        (?) <u>Uria</u> sp.
        (?) <u>Cepphus</u> sp.
        (?) <u>Aethia</u> sp.
        Fraterculini, gen. and sp. indet.
        *<u>Praemancalla wetmorei</u>

Snake Creek, (Laucomer Member)
    Formation: Snake Creek, Laucomer Mbr.
    State: Nebraska, Sioux Co.
    NALMA or Correlative: Late Clarendonian
    Radiometric Dates: none
    General References: Skinner et al. 1977
    Avian References: Wetmore 1923, 1928a
    Avian Taxa: Cracidae
            *<u>Boreortalis phengites</u>
        Gruidae
            Balearicinae
            "<u>Grus canadensis</u>"

52

Comments: This list includes unstudied material in the Frick Collections.

Leisure World
    Formation: Monterey
    State: California, Orange Co.
    NALMA or Correlative: Clarendonian
    Radiometric Dates: none
    General References: Savage and Barnes 1972
    Avian References: Howard 1966b, 1968
    Avian Taxa: Diomedeidae
            Diomedea, 2 sp.
        Procellariidae
            *Puffinus calhouni
            Puffinus priscus
            Puffinus sp.
            *Fulmarus hammeri
        Sulidae
            Microsula sp.
            Morus lompocanus?
            (?) Miosula sp.
            Sulidae sp.
        Pelagornithidae
            Osteodontornis orri?
        Anatidae
            Presbychen abavus
            Anserinae sp.
        Alcidae
            Alca sp.
            Cerorhinca sp.
            *Aethia rossmoori
            *Alcodes ulnulus
            *Praemancalla lagunensis
            Alcidae sp.

Comments: This locality may possibly be older than Clarendonian.

Little Beaver A
    Formation: Ash Hollow (Ogallala Group)
    State: Nebraska, Cherry Co.
    NALMA or Correlative: Clarendonian
    Radiometric Dates: none
    General References: Webb 1969a
    Avian References: A. Miller and Sibley 1941
    Avian Taxa: Laridae
            *Gaviota niobrara

Lomita
    Formation: Monterey, Valmonte Mbr.
    State: California, Los Angeles Co.
    NALMA or Correlative: Clarendonian
    Radiometric Dates: none
    General References: Savage and Barnes 1972
    Avian References: L. Miller 1935, Howard 1958
    Avian Taxa: Procellariidae
            Puffinus diatomicus
        Sulidae
            *Palaeosula stocktoni
            Sula willetti
    Comments: This list includes the San Pedro Breakwater local fauna.

Lompoc
    Formation: Sisquoc
    State: California, Santa Barbara Co.
    NALMA or Correlative: Mohnian (Clarendonian)
    Radiometric Dates: none
    General References: Savage and Barnes 1972, Repenning and Tedford 1977

Avian References: L. Miller 1925, Orr 1940, Howard 1981

Avian Taxa: Procellariidae
 *Puffinus diatomicus
 Sulidae
  *Sula willetti
  *Morus lompocanus
  *Miosula media
 Scolopacidae
  *Limosa vanrossemi
 Alcidae
  *Cerorhinca dubia
  *Uria brodkorbi

Comments: See L. Miller and DeMay (1942) for other references.

Love Bone Bed
 Formation: Alachua
 State: Florida, Alachua Co.
 NALMA or Correlative: Latest Clarendonian
 Radiometric Dates: none
 General References: Webb et al. 1981
 Avian References: Becker 1985a, 1985b, 1985c, 1986d, 1987b
 Avian Taxa: Podicipedidae
  Rollandia sp.
  Tachybaptus sp.
 Phalacrocoracidae
  Phalacrocorax sp.
 Anhingidae
  Anhinga grandis
 Ardeidae
  Ardea sp.
  Egretta sp.
  Ardeola sp.

Ciconiidae
    Mycteria sp.
    Ciconia sp.
Plataleidae
    Plegadis cf. P. pharangites
    Eudociminae, genus and species indet.
Anatidae
    Dendrocygna sp.
    Branta sp.
    Anserinae, genus indet., 2 sp.
    Anas size near A. acuta
    Anas n. sp.
    Anatini, genus indet., 2 sp.
Vulturidae
    *Pliogyps charon
Pandionidae
    *Pandion lovensis
Accipitridae
    Genus indet., 2 sp.
Phasianidae
    Meleagridinae, genus indet.
Rallidae
    Rallus, 2 sp.
    Genus to be described
Gruidae
    Grus, 2 sp.
    Aramornis sp.
Jacanidae
    Jacana farrandi
Scolopacidae
    "Calidris" 3 sp.
    (?) Actitis sp.
    (?) Arenaria sp.
    Genus indet., 2 sp.

        Phoenicopteridae
            Phoenicopterus, 2 sp.
        Tytonidae
            Genus to be described
        Passeriformes
            Suborder indet., 2 sp.

Merritt Dam (Brown Co.)
    Formation: Ash Hollow, Merritt Dam Mbr.
    State: Nebraska, Brown Co.
    NALMA or Correlative: Late Clarendonian
    Radiometric Dates: none in Brown County
    General References: Skinner and Johnson 1984
    Avian References: none
    Avian Taxa: Anatidae
                Phasianidae
    Comments: This list includes unstudied material from the Frick Pratt, Pratt Slide (=Pratt), and East Clayton quarries. See discussion by Skinner and Johnson (1984).

Merritt Dam (Cherry Co.)
    Formation: Ash Hollow, Merritt Dam Mbr.
    State: Nebraska, Cherry Co.
    NALMA or Correlative: Late Clarendonian
    Radiometric Dates: 10.2 ± 0.7 mybp on glass shards of the Davis Ash in type section (Skinner and Johnson 1984).
    General References: Skinner and Johnson 1984
    Avian References: none
    Avian Taxa: Accipitridae
                Strigiformes
    Comments: This list includes unstudied material from the Frick Gallup Gulch and Bear Creek quarries. See discussion by Skinner and Johnson (1984).

Minnechaduza
    Formation: Ash Hollow, Cap Rock Mbr.
    State: Nebraska, Cherry Co.
    NALMA or Correlative: Early Clarendonian
    Radiometric Dates: none
    General References: Webb 1969a, Skinner and Johnson 1984
    Avian References: Webb 1969a
    Avian Taxa: Gruidae
                  Grus sp.

Poison Ivy Quarry
    Formation: Ash Hollow, Cap Rock Mbr.
    State: Nebraska, Antelope Co.
    NALMA or Correlative: Late Clarendonian
    Radiometric Dates: none
    General References: Voorhies and Thomasson 1979
    Avian References: Feduccia, pers. comm. 1986
    Avian Taxa: Accipitridae (2 genera)
               Gruidae
                  Balearicinae
                     Balearica n. sp.
               Lari
                  Sterninae

Pojoaque
    Formation: Tesuque, Pojoaque Mbr.
    State: New Mexico, Santa Fe Co.
    NALMA or Correlative: Late Barstovian and Clarendonian
    Radiometric Dates: 9.4 ± 0.9, 11.4 ± 1.1 mybp, fisson track dates (Izett and Naeser 1981) considered minimum dates by Barghoorn (1981).
    General References: Galusha and Blick 1971

Avian References: none
Avian Taxa: Anatidae
Accipitridae
Falconidae
Phasianidae
  Tetraoninae
Rallidae
Strigidae
Picidae
Passeriformes
Comments: This list includes unstudied material from the Frick Collections from late Barstovian Pojoaque Bluffs and from the Jacona Microfauna Quarry, currently under study by D. Chaney (USNM). The Pojoaque Member overlies the Skull Ridge Member but see Galusha and Blick (1971) for additional comments.

Porter
Formation: Clarendon
State: Texas, Donley Co.
NALMA or Correlative: Early Clarendonian
Radiometric Dates: none
General References: none
Avian References: none
Avian Taxa: Anatidae (2 species)
Comments: This list is based on unstudied material in the Frick Collections.

Ricardo (Restricted)
Formation: Ricardo
State: California, Kern Co.
NALMA or Correlative: Early Clarendonian
Radiometric Dates: none

General References: Merriam 1919, Whistler 1970,
   Tedford et al. in press
Avian References: L. Miller 1930
Avian Taxa: Anatidae
   *Branta howardae
Comments: See L. Miller and DeMay (1942) for
   additional references. See also Dove Springs local
   fauna, above.

Round Mountain
   Formation: "Chamita"
   State: New Mexico, Rio Arriba
   NALMA or Correlative: Early Clarendonian
   Radiometric Dates: none near site
   General References: Galusha and Blick 1971, Tedford
      1981
   Avian References: none
   Avian Taxa: Anatidae
   Comments: This record is based on unstudied material
      in Frick collections from the Round Mountain
      Quarry. Tedford (1981) discusses the stratigraphy
      of this area and gives a preliminary faunal list of
      the mammals from the Round Mountain Quarry.

San Fernando Valley (Studio City)
   Formation: Modelo
   State: California, Los Angeles Co.
   NALMA or Correlative: Clarendonian
   Radiometric Dates: none
   General References: Barnes 1976, Repenning and
      Tedford 1977
   Avian References: Howard 1958
   Avian Taxa: Sulidae
         *Sula pohli
         Sula (?) sp.

Tepusquet Canyon
    Formation: Monterey
    State: California, Santa Barbara Co.
    NALMA or Correlative: Mohnian (Clarendonian)
    Radiometric Dates: none
    General References: Savage and Barnes 1972
    Avian References: Howard 1957a, 1957b, Harrison and Walker 1976
    Avian Taxa: Procellariidae
            Puffinus sp.
           Pelagornithidae
            Osteodontornis orri
           Palaeoscinidae
            Palaeoscinis turdirostris

Wakeeney
    Formation: Ogallala Group
    State: Kansas, Trego Co.
    NALMA or Correlative: Mid to Late Clarendonian
    Radiometric Dates: none
    General References: Wilson 1968
    Avian References: Brodkorb 1962, Feduccia and Wilson 1967
    Avian Taxa: Anatidae
            Anas ogallalae
           Cracidae
            Ortalis affinis
           Picidae
            Pliopicus brodkorbi

Westmoreland State Park
    Formation: Eastover, Claremont Mbr.
    State: Virginia, Westmoreland Co.
    NALMA or Correlative: Clarendonian (?)

Radiometric Dates: none
General References: none
Avian References: Steadman 1980
Avian Taxa: Phasianidae
              cf. *Meleagris*

Xmas--Kat Channel
   Formation: Ash Hollow, Merritt Dam Mbr.
   State: Nebraska, Cherry Co.
   NALMA or Correlative: Late Clarendonian
   Radiometric Dates: none
   General References: Skinner and Johnson 1984
   Avian References: none
   Avian Taxa: Anatidae
              Anatinae
              Phasianidae
   Comments: This list includes unstudied material from the Frick Hans Johnson and *Machairodus* quarries. See discussion by Skinner and Johnson (1984).

Hemphillian

[handwritten note: Yepomera, Chihuahua Mex listed here (p.85)]

Aphelops Draw
Formation: Snake Creek, Johnson Mbr.
State: Nebraska, Sioux Co.
NALMA or Correlative: Early Hemphillian
Radiometric Dates: none
General References: Skinner et al. 1977
Avian References: Wetmore 1923
Avian Taxa: Accipitridae
    *Buteo conterminus

Arnett
Formation: Ogallala Group
State: Oklahoma, Ellis Co.
NALMA or Correlative: Early Hemphillian
Radiometric Dates: none
General References: Schultz 1977
Avian References: none
Avian Taxa: Accipitridae
Comments: This record is based on unstudied material in the Frick Collections. This locality is also known as the Point of Entry local fauna (Tedford et al. in press).

Bone Valley
Formation: Bone Valley
State: Florida, Polk Co.
NALMA or Correlative: Late Hemphillian, but see comments below.
Radiometric Dates: none
General References: MacFadden 1982, MacFadden and Webb 1982, Berta and Morgan 1985, Webb and Waldrop in prep.

Avian References: Wetmore 1943b, Brodkorb 1953a, 1953b, 1953c, 1953d, 1955, 1967, 1970, Warter 1976, Olson and Steadman 1978, Steadman 1980, Becker 1985a, 1985c

Avian Taxa: Gaviidae
    *Gavia palaeodytes
    Gavia concinna
Podicipedidae
    Podilymbus cf. P. podiceps
    Podiceps sp.
    *Pliodytes lanquisti
Diomedeidae
    Diomedea cf. D. anglica
Procellariidae
    Puffinus sp.
Pelecanidae
    Pelecanus sp.
Sulidae
    *Morus peninsularis
    *Sula guano
    *Sula phosphata
Phalacrocoracidae
    *Phalacrocorax wetmorei
    Phalacrocorax cf. P. idahensis
Anhingidae
    Anhinga sp.
Plataleidae
    Eudocimus sp.
Ardeidae
    *Ardea polkensis
    Egretta sp.
Ciconiidae
    Ciconia, 2 sp.
Pandionidae
    Pandion sp.

Accipitridae
    (?) <u>Haliaeetus</u> sp.
    <u>Buteo</u> sp.
    <u>Aquila</u> sp.
    Genus indet., 2 sp.
Anatidae
    Anserinae, genus indet., 2 sp.
    Tadornini, genus indet.
    <u>Aythya</u> sp.
    <u>Oxyura</u> cf. <u>O</u>. <u>dominica</u>
    *<u>Bucephala ossivallis</u>
Phasianidae
    Meleagridinae, cf. <u>Meleagris</u> sp.
Gruidae
    Balearicinae, genus indet.
Rallidae
    <u>Rallus</u> sp.
Scolopacidae
    *<u>Limosa ossivallis</u>
    *<u>Calidris penepusilla</u>
    *<u>Calidris pacis</u>
    <u>Calidris</u> sp.
    (?) <u>Philomachus</u> sp.
Phoenicopteridae
    *<u>Phoenicopterus floridanus</u>
Haematopodidae
    *<u>Haematopus sulcatus</u>
Laridae
    *<u>Larus elmorei</u>
Alcidae
    (?) <u>Pinguinus</u> sp.
    *<u>Australca grandis</u>
    undescribed alcidae, (?) 3 sp.
Strigidae
    <u>Bubo</u> sp.

Comments: *Ortalis* was reported in error from the Bone Valley (Brodkorb 1970). "Bone Valley" refers to a collection of local faunas. The Lower Bone Valley fauna is usually considered early Hemphillian and the Upper Bone Valley fauna is late Hemphillian. At least two earlier local faunas are known (Barstovian and Clarendonian age), but to date none of the earlier local faunas have produced fossil birds. I know of no evidence that would indicate the fossil birds are from a stratigraphic level in the Bone Valley Formation other than the one producing the majority of the fossil mammals -- the late Hemphillian Upper Bone Valley local fauna.

Box T
  Formation: Hemphill Beds
  State: Texas, Lipscomb Co.
  NALMA or Correlative: Late Hemphillian
  Radiometric Dates: none
  General References: Schultz 1977
  Avian References: none
  Avian Taxa: Anatidae
              Anserinae
  Comments: This record is based on unstudied material in the Frick Collections.

Cambridge
  Formation: "Kimball" (see comments below)
  State: Nebraska, Frontier Co.
  NALMA or Correlative: Early Hemphillian
  Radiometric Dates: none near site
  General References: Schultz et al. 1970, Breyer 1981, Voorhies 1984
  Avian References: Short 1969, Martin and Tate 1970, Martin and Mengel 1975, Martin 1975a, Steadman 1980

Avian Taxa: Anhingidae
Anhinga grandis
Anatidae
Paracygnus plattensis
Accipitridae
Spizaetus schultzi
Phasianidae
Proagriocharis kimballensis
Comments: The vertebrate fauna from this locality (UNSM Loc. Ft. 40.) was one of several originally used to characterize the "Kimballian Land Mammal Age." Most workers now regard this "age" as artificial (Breyer 1981) and the "Kimball Formation" as part of the Ash Hollow Formation (Tedford et al. in press).

Capistrano Beach
    Formation: Capistrano
    State: California, Orange Co.
    NALMA or Correlative: Early Hemphillian
    Radiometric Dates: none
    General References: Savage and Barnes 1972, Fife 1974
    Avian References: L. Miller 1951, Howard 1978
    Avian Taxa: Oceanitidae
        Oceanodroma hubbsi

Carlin High Level Quarry
    Formation: unnamed
    State: Nevada, Elko Co.
    NALMA or Correlative: Late Hemphillian
    Radiometric Dates: none
    General References: none
    Avian References: none
    Avian Taxa: Anatidae
        Anserinae

Comments: This record is based on unstudied material in the Frick Collections.

Castle Creek
    Formation: "Chalk Hills" (see comments below)
    State: Idaho, Owyhee Co.
    NALMA or Correlative: "Hemphillian"
    Radiometric Dates: none
    General References: Ekren et al. 1978, Kimmel 1979
    Avian References: Marsh 1870, Shufeldt 1915
    Avian Taxa: Phalacrocoracidae
              *Phalacrocorax idahensis
    Comments: The type locality of this species was given as "Castle Creek" by Marsh (1870). Sediments exposed along Castle Creek include those from the Chalk Hills, Glenns Ferry, and Bruneau formations, in addition to unnamed deposits of Quaternary alluvium. These sediments range in age from late Miocene through late Pleistocene (Ekren et al. 1978).

Cedros Island
    Formation: Almejas, lower member
    State: Cedros Is., Baja California, Mexico
    NALMA or Correlative: Hemphillian
    Radiometric Dates: none
    General References: Repenning and Tedford 1977, Barnes 1973
    Avian References: Howard 1971
    Avian Taxa: Procellariidae
             *Puffinus tedfordi
             Puffinus sp.
          Sulidae
             Morus sp.
          Phoenicopteridae

                (?) _Megapaloelodus opsigonus_
            Alcidae
                *_Cerorhinca minor_
                Brachyramphus (?) sp.
                *_Mancalla cedrosensis_

Clifton Country Club
    Formation: unreported
    State: Arizona, Graham Co.
    NALMA or Correlative: Hemphillian
    Radiometric Dates: none
    General References: none
    Avian References: Steadman 1980
    Avian Taxa: Phasianidae
                Meleagridinae
                    Genus and species indeterminate

Coffee Ranch
    Formation: Hemphill Beds
    State: Texas, Hemphill Co.
    NALMA or Correlative: Late Hemphillian
    Radiometric Dates: 4.7 ± 0.8, 5.3 ± 0.4, and 6.6 ± 0.4 mybp all immediately above quarry
    General References: Reed and Longnecker 1932, Schultz 1977, Dalquest 1983 and references therein.
    Avian References: Compton 1934, Wetmore 1944
    Avian Taxa: Anatidae
                Anas crecca
    Comments: This late Hemphillian record of a living species of teal requires verification.

Corona Del Mar
    Formation: uncertain, see comments below
    State: California, Orange Co.
    NALMA or Correlative: Hemphillian

Radiometric Dates: none
General References: none
Avian References: Lucas 1901a, Lucas 1901b, Howard 1949, L. Miller and Howard 1949
Avian Taxa: Procellariidae
   *Puffinus felthami
  Alcidae
   *Mancalla californiensis
Comments: This local fauna is not from the Repetto Formation. The presence of Mancalla indicates a Hemphillian or Blancan age and possibly these fossils originated in the Capistrano Formation (Howard, pers. comm.).

Edson
 Formation: Ogallala Group
 State: Kansas, Sherman Co.
 NALMA or Correlative: Late Hemphillian
 Radiometric Dates: none
 General References: Martin 1928, Harrison 1981
 Avian References: Wetmore and Martin 1930, Wetmore 1937
 Avian Taxa: Podicipedidae
   Podiceps nigricollis
  Vulturidae
  Gruidae
   *Grus nannodes
  Rallidae
  Scolopacidae
  Corvidae

Etchegoin
 Formation: Etchegoin
 State: California, Monterey Co.
 NALMA or Correlative: Late Hemphillian

Radiometric Dates: none
General References: Merriam 1915, Nomland 1916, Repenning 1976, Repenning and Tedford 1977,
Avian References: Wetmore 1940
Avian Taxa: Gaviidae
          *Gavia concinna
Comments: The Etchegoin Formation grades laterally into Kern River Formation. Synonymous with "Sweetwater Canyon" (L. Miller and DeMay 1942).

Feltz Ranch
    Formation: Ash Hollow, Merritt Dam Mbr.
    State: Nebraska, Keith Co.
    NALMA or Correlative: Early Hemphillian
    Radiometric Dates: none
    General References: Hesse 1935, Skinner and Johnson 1984
    Avian References: Compton 1935b
    Avian Taxa: Anatidae
              Anserinae
    Comments: See discussion by Skinner and Johnson (1984) and Tedford et al. (in press) concerning this locality.

Haile VI
    Formation: unnamed
    State: Florida, Alachua Co.
    NALMA or Correlative: Hemphillian
    Radiometric Dates: none
    General References: Webb 1966
    Avian References: Brodkorb 1963a, Steadman 1981
    Avian Taxa: Passeriformes
              *"Palaeostruthus" eurius
    Comments: Brodkorb (1963a) originally described this passerine as a new species, but Steadman (1981)

considers the specimen to be indeterminate at both
the family and the generic level and certainly not
referable to Palaeostruthus, which he synonymized
with the living genus Ammodramus.

Haile XIXA
    Formation: unnamed
    State: Florida, Alachua Co.
    NALMA or Correlative: Hemphillian
    Radiometric Dates: none
    General References: Becker 1985c
    Avian References: Becker 1985c, 1986c, 1987b
    Avian Taxa: Anhingidae
            Anhinga grandis
        Phalacrocoracidae
            Phalacrocorax sp.
        Anatidae
            Anatinae, genus indet.
        Momotidae
            Genus and species indet.

Hernandez School House
    Formation: Chamita
    State: New Mexico, Rio Arriba Co.
    NALMA or Correlative: Hemphillian
    Radiometric Dates: none
    General References: none
    Avian References: none
    Avian Taxa: Accipitridae
    Comments: This record is based on unpublished
        material in the Frick collections.

Higgins
    Formation: Ogallala Group
    State: Texas, Lipscomb Co.

NALMA or Correlative: Early Hemphillian
Radiometric Dates: none
General References: Schultz 1977 and references therein.
Avian References: none
Avian Taxa: Ciconiidae
Phasianidae
Comments: This list is based on unstudied material in the Frick Collections.

Hill Point
Formation: Goodnight Beds
State: Texas, Armstrong Co.
NALMA or Correlative: Late Hemphillian
Radiometric Dates: none
General References: Schultz 1977
Avian References: none
Avian Taxa: Corvidae
Comments: This record is based on unstudied material in the Frick Collections.

J. Swayze Quarry
Formation: Ogallala Group
State: Kansas, Clark Co.
NALMA or Correlative: Early Hemphillian
Radiometric Dates: none
General References: none
Avian References: none
Avian Taxa: Vulturidae
Comments: This record is based on unstudied material in the Frick Collections.

Juntura Basin
Formation: Drewsey
State: Oregon, Malheur Co.

NALMA or Correlative: Early Hemphillian
Radiometric Dates: 9.4 ± 0.6 mybp from a welded tuff from the lowest unit of the Drewsey Formation (below local fauna).
General References: Shotwell 1963
Avian References: Brodkorb 1961
Avian Taxa: Phalacrocoracidae
    *Phalacrocorax leptopus
    Ciconiidae
    Accipitridae
    Neophrontops dakotensis
Comments: This local fauna is from University of Oregon locality 2360 and is stratigraphically superimposed over the Juntura Formation containing the Clarendonian Black Butte local fauna.

Lawrence Canyon
    Formation: San Mateo
    State: California, San Diego Co.
    NALMA or Correlative: Hemphillian
    Radiometric Dates: none
    General References: Barnes et al. 1981
    Avian References: Barnes et al. 1981, Howard 1982
    Avian Taxa: Sulidae
        Genus and sp. indet.
    Accipitridae
        Genus and sp. indet
    Falconidae
        (?) Falco sp.
    Alcidae
        Cepphus sp.
        Mancalla milleri
        Mancalla diegensis
        Mancalla cf. M. cedrosensis
        Mancalla sp. indet.

Comments: This local fauna is younger than the San Luis Rey River local fauna. The San Mateo Formation may be a facies of the Capistrano Formation (Howard 1982).

Lee Creek
   Formation: Yorktown
   State: North Carolina, Beaufort Co.
   NALMA or Correlative: Hemphillian; late zone N18 and zone N19
   Radiometric Dates: 4.5 ± 0.2 mybp glauconite potassium-argon date on the Yorktown Formation in Virginia (Blackwelder 1981).
   General References: Ray 1976, 1983, Baum and Wheeler 1977, Repenning and Tedford 1977, Tedford and Hunter 1984
   Avian References: Olson 1977a, 1981b, Olson and Steadman 1978, Olson pers. comm., 1983, 1986
   Avian Taxa: Gaviidae
                     Gavia, 3 sp.
                 Podicipedidae
                     Podiceps ? sp.
                 Diomedeidae
                     Diomedea, 4 sp.
                 Procellariidae
                     Puffinus, 4 sp.
                 Pelecanidae
                     Pelecanus sp.
                 Pelagornithidae
                     Pseudodontornis sp.
                 Incertae Sedis
                     Palaeochenoides sp.
                 Phalacrocoracidae
                     Phalacrocorax, 2 sp.

Sulidae
   Sulid, 5 sp.
   Microsula avita
Plataleidae
   Eudocimus sp.
Ciconiidae
Anatidae
   Cygninae
   Anserinae
   Anatinae and Aythyinae, 4 sp. minimum
Pandionidae
   Pandion sp.
Phasianidae
   Meleagridinae
Rallidae
Gruidae
   Grus sp.
Haematopodidae
   Haematopus sp.
Phoenicopteridae
Stercorariidae
   Catharacta sp.
   Stercorarius sp.
Laridae
   Laridae, 5 sp.
Alcidae
   Australca, at least 2 sp.
   **Pinguinus alfrednewtoni*
   Miocepphus sp.
   Alle sp.
   (?) *Cyclorrhynchus* sp.
   Fratercula, 3 sp.
Cuculidae
Columbidae

                Corvidae
                  Corvus sp.

Comments: The above list is still incomplete. Fossil birds at Lee Creek occur in two stratigraphic levels. Most specimens, and the majority of species, are from the lower part of the Yorktown Formation and are from the same stratigraphic levels as the land mammals. A few species, listed above, (*Pseudodontornis*, *Microsula*, and *Miocepphus*, and possibly others) are from the underlying Pungo River Formation. These taxa are similar to those from the Calvert Formation. Terrestrial mammals from the basal Yorktown, especially the canid *Osteoborus* cf. *O. dudleyi*, indicates a correlation with the local fauna of late Hemphillian age from the upper part of the Bone Valley Formation (Tedford and Hunter 1984).

Long Island
    Formation: Republican River
    State: Kansas, Phillips Co.
    NALMA or Correlative: Early Hemphillian
    Radiometric Dates: none
    General References: Bennett 1984, Skinner et al. 1977
    Avian References: Shufeldt 1913, Steadman 1981
    Avian Taxa: Accipitridae
                *Proictinia gilmorei
            Gruidae
                Balearicinae
            Emberizidae
                *Ammodramus hatcheri

"Loup Fork"
    Formation: (?) Ogallala Group
    State: Nebraska

NALMA or Correlative: (?) Hemphillian
Radiometric Dates: none
General References: Skinner and Johnson 1984
Avian References: Marsh 1871, Shufeldt 1915, Wetmore 1926b, Brodkorb 1964b
Avian Taxa: Accipitridae
*Buteo dananus
Comments: This species, from a "Pliocene bluff on the Loup Fork river" (Marsh 1871: 126), is of uncertain stratigraphic position. There has been little agreement concerning its generic placement: Marsh (1871) described this specimen as a species of Aquila; Shufeldt (1915) considered it definitely to be accipitrid, noted similarities between it and the "vulturine accipitrids" (i.e. Gypaetus), then concluded that it was not complete enough to make a positive identification; Wetmore (1926) moved it to Geranoaetus; and Brodkorb (1964b) listed it as a species of Buteo. I have arbitrarily followed Brodkorb's (1964b) generic placement of this species.

Manatee County Dam
    Formation: Bone Valley
    State: Florida, Manatee Co.
    NALMA or Correlative: Mid Hemphillian
    Radiometric Dates: none
    General References: Webb and Tessman 1968
    Avian References: Webb and Tessman 1968, Becker 1985c
    Avian Taxa: Phalacrocoracidae
            Phalacrocorax cf. P. wetmorei

McGehee
    Formation: Alachua
    State: Florida, Alachua Co.

NALMA or Correlative: Early Hemphillian
Radiometric Dates: none
General References: Webb 1964, Webb 1969b, Hirschfeld and Webb 1968, Becker 1985c, 1987b
Avian References: Brodkorb 1963a, Olson 1976, Becker 1985c
Avian Taxa: Podicipedidae
  Rollandia sp.
  Tachybaptus sp.
  Podiceps sp.
 Phalacrocoracidae
  Phalacrocorax sp.
 Anhingidae
  Anhinga grandis
 Ciconiidae
  Mycteria sp.
 Ardeidae
  *Nycticorax fidens
 Anatidae
  Anas sp. (smaller than A. acuta)
  Anas sp. (size near A. acuta)
 Rallidae
  Undescribed genus
 Jacanidae
  *Jacana farrandi
 Scolopacidae
  *Calidris rayi
  Calidris, 2 sp.
 Phoenicopteridae
  Phoenicopterus sp.

McKay Reservoir
 Formation: Dalles
 State: Oregon, Umatilla Co.
 NALMA or Correlative: Late Hemphillian

Radiometric Dates: none
General References: Shotwell 1955, 1956, Newcomb 1971
Avian References: Brodkorb 1958a
Avian Taxa: Anatidae
        Anas bunkeri
    Phasianidae
        *Lophortyx shotwelli
    Scolopacidae
        *Bartramia umatilla

Mixson
    Formation: Alachua
    State: Florida, Levy Co.
    NALMA or Correlative: Early Hemphillian
    Radiometric Dates: none
    General References: Leidy and Lucas 1896, Simpson 1930, Webb 1969b, Becker 1985c
    Avian References: Becker 1985c
    Avian Taxa: Podicipedidae
        Rollandia sp.
        Podilymbus sp.
    Ciconiidae
        Ciconia sp.

Optima or Guymon
    Formation: Ogallala Group
    State: Oklahoma, Texas Co.
    NALMA or Correlative: Late Hemphillian
    Radiometric Dates: none
    General References: Tedford et al. in press
    Avian References: none
    Avian Taxa: Rallidae
    Comments: This record is based on unstudied material in the Frick Collections.

Pozo Creek
  Formation: Kern River
  State: California, Kern Co.
  NALMA or Correlative: Early to Mid Hemphillian
  Radiometric Dates: none
  General References: Repenning and Tedford 1977,
    Tedford et al. in press
  Avian References: Miller, L. 1931
  Avian Taxa: Vulturidae
          *Sarcoramphus kernensis
        Accipitridae
          (?) Parabuteo sp.
  Comments: This locality is synonymous with Kern River
    Divide (L. Miller and DeMay 1942).

San Luis Rey River
  Formation: San Mateo
  State: California, San Diego Co.
  NALMA or Correlative: Early Hemphillian
  Radiometric Dates: none
  General References: Barnes et al. 1981
  Avian References: Barnes et al. 1981, Howard 1982
  Avian Taxa: Gaviidae
          Gavia sp.
        Diomedeidae
          Diomedea sp.
        Alcidae
          *Uria paleohesperis
          *Cepphus olsoni
          (?) Aethia sp.
          Praemancalla cf. P. wetmorei
  Comments: See comments under Lawrence Canyon local
    fauna.

Star Valley
    Formation: sediments correlative to Big Island Fm.
    State: Idaho, Owyhee Co.
    NALMA or Correlative: Early Hemphillian
    Radiometric Dates: none
    General References: Becker 1980, Coats 1985
    Avian References: Becker 1980, 1987a
    Avian Taxa: Falconidae
            (?) *Falco* sp.
          Columbidae
            Zenaida sp.
          Passeriformes

Tarboro
    Formation: Yorktown
    State: North Carolina, Edgecombe Co.
    NALMA or Correlative: Latest Hemphillian
    Radiometric Dates: none
    General References: Ray 1983
    Avian References: Marsh 1870, Olson and Gillette 1978, Olson 1985b
    Avian Taxa: Alcidae
          Australca antiqua
    Comments: See Olson and Gillette (1978) and Olson (1985b) for a discussion of the age and systematic position of this species.

Wickieup
    Formation: Big Sandy
    State: Arizona, Mohave Co.
    NALMA or Correlative: Latest Hemphillian
    Radiometric Dates: 5.5 ± 0.2 mybp for a mean of three fisson track dates (MacFadden et al. 1979)
    General References: Sheppard and Gude 1972, MacFadden et al. 1979

Avian References: Wetmore 1943a, 1957, Bickart 1986
Avian Taxa: Podicipedidae
        <u>Podiceps</u> sp.
        <u>Podilymbus</u> sp.
    Ciconiidae
        <u>Ciconia</u> sp.
    Anatidae
        <u>Cygnus</u> new sp.
        <u>Anser</u>, 2 new sp.
        <u>Branta</u>, new sp.
        cf. <u>Anabernicula</u> sp.
        Anserinae, genus indet., 2 sp.
        <u>Anas</u>, 3 sp.
        Anatinae, genus and sp. indet.
    Accipitridae
        <u>Aquila</u>, 2 sp.
        <u>Buteo</u>, 2 sp.
        <u>Neophrontops</u> sp.
        <u>Circus</u> sp.
    Gruidae
        <u>Grus haydeni</u>
    Rallidae
        *<u>Rallus phillipsi</u>
        <u>Rallus</u> sp.
        <u>Coturnicops</u> sp.
        Genus and sp. indet.
    Phoenicopteridae
        <u>Phoenicopterus</u> sp.
    Recurvirostridae
        <u>Himantopus</u> new sp.
        <u>Recurvirostra</u> sp.
    Charadriidae
        <u>Charadrius</u> sp.

Scolopacidae
cf. *Limosa* sp.
cf. *Tringa*, 2 sp.
cf. *Calidris*, 3 sp.
Laridae
cf. *Larus* sp.
Columbidae
cf. *Zenaida prior*
Corvidae
Corvus new sp.

Comments: Most of this large avifauna is housed in the Frick collections of the American Museum of Natural History, with smaller collections at the Los Angeles County Museum (including the CIT collections), the University of Arizona, the Museum of Northern Arizona, the National Museum of Natural History, Smithsonian Institution and West Texas University.

Withlacoochee River 4A
    Formation: Alachua
    State: Florida, Marion Co.
    NALMA or Correlative: Mid Hemphillian
    Radiometric Dates: none
    General References: Hirschfeld and Webb 1968, Webb 1969b, Becker 1985b, 1985c
    Avian References: Becker 1985b, 1985c
    Avian Taxa: Ardeidae
                Egretta subfluvia
            Accipitridae
                Buteo sp., size near *B. jamaicensis*

Wray
- Formation: Ogallala Group, (Ash Hollow Fm. ?)
- State: Colorado, Yuma Co.
- NALMA or Correlative: Early Hemphillian
- Radiometric Dates: none
- General References: Tedford et al. in press.
- Avian References: none
- Avian Taxa: Anatidae
- Comments: This record is based on unstudied material in the Frick Collections.

XZ Bar
- Formation: Snake Creek, upper part of the Johnson Member
- State: Nebraska, Sioux Co.
- NALMA or Correlative: Late Hemphillian
- Radiometric Dates: none
- General References: Skinner et al. 1977
- Avian References: none
- Avian Taxa: Phasianidae
- Comments: This record includes unstudied material from Pliohippus Draw in the Frick collections.

Yepomera (=Rincon)
- Formation: unnamed
- State: Chihuahua, Mexico
- NALMA or Correlative: Latest Hemphillian
- Radiometric Dates: none
- General References: Lance 1950, Jacobs and Lindsay 1981
- Avian References: L. Miller 1944a, Howard 1966a, Steadman and McKitrick 1982

Avian Taxa: Anatidae
        Eremochen cf. E. russelli
        Anas bunkeri
        *Wasonaka yepomerae
        Oxyura sp.
    Scolopacidae
        Calidris sp.
    Phoenicopteridae
        *Phoenicopterus stocki
    Emberizidae
        cf. Passerina sp.

handwritten note at top: Vallecito Creek p. 104 listed in this age

Blancan

Beck Ranch
  Formation: unnamed
  State: Texas, Scurry Co.
  NALMA or Correlative: early Blancan
  Radiometric Dates: none
  General References: Dalquest and Donovan 1973
  Avian References: Brodkorb 1971b
  Avian Taxa: Picidae
           *Campephilus dalquesti

Benson
  Formation: Saint David, middle member
  State: Arizona, Cochise Co.
  NALMA or Correlative: Early Blancan
  Radiometric Dates: Ash underlying the Post Ranch fossil bed has been dated at 3.1 ± 0.7 mybp (Lindsay et al. 1975)
  General References: Gazin 1942, Lindsay et al. 1975
  Avian References: Wetmore 1924, 1944, Holman 1961
  Avian Taxa: Podicipedidae
           Podiceps sp.
         Anatidae
           *Dendrocygna eversa
           *Anabernicula minuscula
           Anas bunkeri
           Anatidae, indet.
         Phasianidae
           Meleagris cf. M. progenes
           Colinus sp.
         Rallidae
           Gallinula sp.
         Scolopacidae
           *Micropalama hesternus

    Corvidae
     Corvus sp.
    Emberizidae
     Junco sp.
    Fringillidae, indet.
Comments: See Lindsay et al. (1975) for paleomagnetics and Kurtén and Anderson (1980) for a short summary of this local fauna.

Blanco
  Formation: Blanco
  State: Texas, Crosby Co.
  NALMA or Correlative: Blancan
  Radiometric Dates: The absolute age of this fauna is debated. Glass shards from the Blanco Ash (above fossiliferous zone) have been dated on fission tracks at 2.8 ± 0.3 mybp (Schultz 1977) and at 1.4 ± 0.2 mybp (Izett et al. 1972). Paleomagnetic studies show the entire fossiliferous section to be reversely magnetized and Lindsay et al. (1975) place it in the Matuyama interval (younger than 2.4 mybp).
  General References: Schultz 1977 and references therein
  Avian References: Shufeldt 1892
  Avian Taxa: Rallidae
     Creccoides osbornii
  Comments: Olson (1977b) was not able to relocate the holotype of this species and suggested it may not belong to the Rallidae.

Broadwater
  Formation: Broadwater, Lisco Mbr.
  State: Nebraska, Morrill Co.
  NALMA or Correlative: Blancan

Radiometric Dates: none
General References: Schultz and Stout 1948; Martin and Mengel 1980
Avian References: Martin and Mengel 1980
Avian Taxa: Anatidae
          *Anser thompsoni
Comments: Kurtén and Anderson (1980) give a short summary of this local fauna. From another locality in the Lisco Mbr., which may be correlative to the above local fauna, D. Chaney (pers. comm.) has recovered a small avifauna that includes accipitrids, gruids, and corvids.

Buckhorn
   Formation: Gila Group
   State: New Mexico, Grant Co.
   NALMA or Correlative: Blancan
   Radiometric Dates: none
   General References: Tedford 1981
   Avian References: Steadman 1980
   Avian Taxa: Anatidae
            Phasianidae
              cf. Meleagris
   Comments: This record includes unstudied material in the AMNH Frick Collections.

Cita Canyon
   Formation: Blanco
   State: Texas, Randall Co.
   NALMA or Correlative: late Blancan
   Radiometric Dates: latest Gauss (about 2.5 mybp)
   General References: Johnstone and Savage 1955, Lindsay et al. 1975
   Avian References: L. Miller and Johnstone 1937, A. Miller and Bowman 1956a, Olson 1981a

Avian Taxa: Plataleidae
                     *Plegadis pharangites
                  Ardeiformes
                     Genus indet.
                  Anatidae
                     Anas sp.
                  Phasianidae
                     Meleagridinae
                        *Meleagris leopoldi
    Comments: Kurtén and Anderson (1980) provide a short summary of this local fauna.

Dry Creek
    Formation: unnamed
    State: Oregon, Malheur Co.
    NALMA or Correlative: Blancan
    Radiometric Dates: none
    General References: L. Miller 1944b
    Avian References: L. Miller 1944b
    Avian Taxa: Pelecanidae
                     Pelecanus erythrorhynchos (?)
                  Phalacrocoracidae
                     Phalacrocorax auritus
                  Anatidae
                     goose, indet.
                  Accipitridae
                     eagle, indet.

Dry Mountain (111 Ranch)
    Formation: Gila Group
    State: Arizona, Graham Co.
    NALMA or Correlative: Late Blancan
    Radiometric Dates: 2.32 ± 0.15 mybp
    General References: Galusha et al. 1984
    Avian References: none

Avian Taxa: Anatidae
              Gruidae
              Phoenicopteridae
Comments: This list includes unstudied material in the AMNH Frick collections and additional material is in the University of Arizona collections.

Duncan
   Formation: Gila Group
   State: Arizona, Greenly Co.
   NALMA or Correlative: Late Blancan
   Radiometric Dates: none
   General References: Tedford 1981
   Avian References: none
   Avian Taxa: Anatidae
                Anserinae
                Anatinae
                Phoenicopteridae
                Accipitridae
   Comments: This list includes unstudied material in the AMNH Frick collections. See Tedford (1981) for a discussion of the biochronology and stratigraphy of the late Cenozoic basins of New Mexico and Arizona.

Flat Iron Butte
   Formation: Glenns Ferry
   State: Idaho, Owyhee Co.
   NALMA or Correlative: Blancan
   Radiometric Dates: about 2.8 to 2.9 mybp based on paleomagnetic correlation.
   General References: Conrad 1980
   Avian References: L. Miller 1944b, Becker in prep.

Avian Taxa: Podicipedidae
   Genera and spp. to be det.
Phalacrocoracidae
   Phalacrocorax sp.
Ciconiidae
   Ciconia maltha
Anatidae
   Anserinae, genus indet.
   Cygnus sp.
   Anatinae, genera and spp. to be det.

Fox Canyon
   Formation: Rexroad
   State: Kansas, Meade Co.
   NALMA or Correlative: Blancan
   Radiometric Dates: about 3.3 to 3.5 mybp based on paleomagnetic correlations.
   General References: Hibbard 1950, Lindsay et al. 1975
   Avian References: Tordoff 1951, Ford 1966, Feduccia 1967a, 1968, Murray 1967, Feduccia and Ford 1970
   Avian Taxa: Podicipedidae
      Pliolymbus baryosteus
      Podiceps discors
   Plataleidae
      Mesembrinibis (?)
      Eudocimus sp.
   Accipitridae
      Buteo sp.
   Falconidae
      Falco sp.
   Phasianidae
      Colinus hibbardi
   Rallidae
   Strigidae
      Otus cf. *asio*

Speotyto megalopeza
Asio sp.
Hirundinidae
Hirundo aprica
Comments: Kurtén and Anderson (1980) provide a summary of the fossil vertebrates from the Rexroad Formation.

Grand View
  Formation: Glenns Ferry
  State: Idaho, Owyhee Co.
  NALMA or Correlative: Blancan
  Radiometric Dates: none
  General References: Shotwell 1970, Conrad 1980
  Avian References: L. Miller 1944b, Becker in prep.
  Avian Taxa: Gruidae
          Grus americana
      Anatidae
          Genera and sp. to be determined.
      Accipitridae
      Rallidae

Hagerman
  Formation: Glenns Ferry
  State: Idaho, Owyhee, Twin Falls and Elmore Co.
  NALMA or Correlative: Blancan
  Radiometric Dates: 3.5 mybp, Evernden et al. 1964
  General References: Zakrzewski 1969, Bjork 1970
  Avian References: Wetmore 1933b, A. Miller 1948, Brodkorb 1958b, Collins 1964, Feduccia 1967b, 1968, 1974, 1975, Ford and Murray 1967, Moseley and Feduccia 1975, Murray 1967, 1970
  Avian Taxa: Podicipedidae
          Pliolymbus baryosteus
          Podiceps discors

Podilymbus majusculus
Aechmophorus elasson
Pelecanidae
    Pelecanus halieus
Phalacrocoracidae
    Phalacrocorax auritus
    Phalacrocorax idahensis
    Phalacrocorax macer
Ardeidae
    Nycticorax sp.
    Egretta sp.
    Genus and sp. indet.
Plataleidae
Ciconiidae
    Ciconia maltha
Anatidae
    Olor hibbardi
    Anser pressus
    Anas platyrhynchos
    Anas bunkeri
    Anas sp.
    Bucephala fossilis
Accipitridae
    Neophrontops slaughteri
Gruidae
    Grus americana
Rallidae
    Rallus prenticei
    Rallus lacustris
    Rallus elegans-longirostris group
    Coturnicops avita
    Gallinula sp.
Strigidae
    Speotyto megalopeza
    Asio brevipes

>   Genus and sp. indet. (near O. asio
>   in size and morphology)
>   Passeriformes
> Comments: This list is from Feduccia (1975) and
>   original references. The identity of many of the
>   reported taxa is in need of further verification
>   and refinement.

Haile XVA
    Formation: unnamed
    State: Florida, Alachua Co.
    NALMA or Correlative: Blancan
    Radiometric Dates: none
    General References: Webb 1974, Robertson 1976,
    Avian References: Campbell 1976, Steadman 1980
    Avian Taxa: Podicipedidae
                    Podilymbus podiceps
                Ardeidae
                    Ardea alba
                    Egretta sp.
                    *Ardeola validipes
                Anatidae
                    Anas crecca
                Phasianidae
                    Colinus cf. C. suilium
                    Meleagris sp.

La Goleta
    Formation: Goleta
    State: Michoacan, Mexico
    NALMA or Correlative: Early Blancan
    Radiometric Dates: none
    General References: Repenning 1962, Arellano and
        Azcon 1949

Avian References: Howard 1965
Avian Taxa: Phalacrocoracidae
Phalacrocorax goletensis

Matthews Wash
    Formation: Gila Group
    State: Arizona, Graham Co.
    NALMA or Correlative: Late Blancan
    Radiometric Dates: none
    General References: none
    Avian References: none
    Avian Taxa: Anatidae
                 Gruidae
                 Phoenicopteridae
    Comments: This list includes unstudied material in the AMNH Frick collections.

Ninefoot Rapids
    Formation: Glenns Ferry
    State: Idaho, Owyhee Co.
    NALMA or Correlative: Blancan
    Radiometric Dates: none
    General References: Conrad 1980
    Avian References: Becker in prep.
    Avian Taxa: Anatidae
                 Genera and spp. to be det.

Oreana
    Formation: Glenns Ferry
    State: Idaho, Owyhee Co.
    NALMA or Correlative: Blancan
    Radiometric Dates: none
    General References: Conrad 1980
    Avian References: Becker 1986e

Avian Taxa: Phalacrocoracidae
    Phalacrocorax idahensis
  Pelecanidae
    Pelecanus cf. *P. halieus*
  Anatidae
  Anatinae
  Strigidae
    Otus near *O. asio*
    Otus near *O. flammeolus*
  Picidae
    Colaptes sp.

Palo Duro Falls
  Formation: (?) Blanco
  State: Texas, Randall Co.
  NALMA or Correlative: (?) Blancan
  Radiometric Dates: none
  General References: Johnson and Savage 1955, Lindsay et al. 1975
  Avian References: A. Miller and Bowman 1956b, Brodkorb 1972
  Avian Taxa: Corvidae
    Protocitta ajax
  Comments: The holotype of *P. ajax* was originally identified by A. Miller and Bowman (1956b) as representing the living species *Pica pica*. Both Miller and Bowman (1956b) and Johnson and Savage (1955) considered this locality as early Pleistocene (i.e. post-Blancan) in age. Brodkorb (1972) restudied this specimen and described it as a new species of *Protocitta*. He also maintained that this locality was more likely to be Blancan in age, without explicitly stating why.

Panaca Valley, Quarry #2
   Formation: Panaca
   State: Nevada, Lincoln Co.
   NALMA or Correlative: Blancan
   Radiometric Dates: none
   General References: May 1981
   Avian References: none
   Avian Taxa: Anatidae
                   Phasianidae
                   Passeriformes
   Comments: These taxa are based on unstudied material
      in the Frick Collections.

Rexroad
   Formation: Rexroad
   State: Kansas, Meade Co.
   NALMA or Correlative: Blancan
   Radiometric Dates: about 3.3 mybp based on
      paleomagnetic correlations.
   General References: Hibbard 1970, Lindsay et al. 1975
   Avian References: Wetmore 1944, Tordoff 1951, 1959,
      Brodkorb 1964c, 1967, 1969, 1972, Collins 1964,
      Ford 1966, Feduccia 1968, Feduccia and Ford 1970,
      Mosley and Feduccia 1975, Steadman 1980
   Avian Taxa: Podicipedidae
                   Pliolymbus baryosteus
                   Podilymbus majusculus
                   Podiceps (?) sp.
                Ardeidae
                   *Botaurus hibbardi
                   Egretta sp. (small)
                   Egretta sp. (medium)
                Plataleidae
                   Plegadis pharangites
                   Eudocimus sp.

98

Mesembrinibis sp. (?)
   Phimosus sp. (?)
Anatidae
   *Anas bunkeri
   Bucephala albeola (?)
   Anatidae, 5 species, not identified
Vulturidae
   *Pliogyps fisheri
Accipitridae
   Accipiter sp.
   Buteo sp.
Falconidae
   Falco sp.
Phasianidae
   *Colinus hibbardi
   *Meleagris progenes
Rallidae
   *Rallus prenticei
   Rallus lacustris
   *Laterallus insignis
   *Gallinula kansarum
Scolopacidae
   Genus and sp. indet.
Laridae
   Sterna sp.
Columbidae
   *Zenaida prior
Strigidae
   Bubo sp.
   Speotyto megalopeza
Psittacidae
   Genus and sp. indet.
Picidae
   Colaptes sp.

        Corvidae
          <u>Protocitta</u> <u>ajax</u>
       Passeriformes, several families
Comments: This list is from Feduccia (1975) and original references. Kurtén and Anderson (1980) provide a summary of the fossil vertebrates from the Rexroad Formation.

St. Petersburg Times Site
   Formation: unnamed
   State: Florida, Pinellas Co.
   NALMA or Correlative: Blancan
   Radiometric Dates: none
   General References: none
   Avian References: Becker, in prep.
   Avian Taxa: Podicipedidae
          Anatidae
            <u>Anas</u>, 2 sp.
          Plataleidae
            <u>Eudocimus</u> sp.
          Rallidae
          Phasianidae

San Diego
   Formation: San Diego
   State: California, San Diego Counties
   NALMA or Correlative: early Blancan
   Radiometric Dates: none
   General References: Barnes 1976
   Avian References: L. Miller 1937, 1956, L. Miller and Howard 1949, Howard 1949, 1970, Brodkorb 1953d, L. Miller and Bowman 1958, Olson 1981b, Chandler 1982, Chandler pers. comm. 1986

Avian Taxa: Gaviidae
    *Gavia howardae
    Gavia sp.
  Podicipedidae
    Podiceps new sp.
    *Podiceps subparvus
    Podiceps sp.
    Aechmophorus cf. A. elasson
  Diomedeidae
    Diomedea, 3 sp.
  Procellariidae
    *Puffinus kanakoffi
    Puffinus new sp.
    Puffinus sp.
  Oceanitidae
    Oceanodroma sp.
  Phalacrocoracidae
    *Phalacrocorax kennelli
    Phalacrocorax new sp.
    Phalacrocorax sp.
  Sulidae
    *Sula humeralis (= Morus)
    *Miosula recentior (= Morus)
    Sula new sp.
    Sula sp.
  Anatidae
    Melanitta sp.
    Mergini
  Laridae
    Rissa new sp.
    Larus sp.
    Sterna sp.
  Alcidae
    *Brachyramphus pliocenus
    Brachyramphus new sp.

        Synthliboramphus new sp.
        Cerorhinca new sp.
        Cerorhinca sp.
        **Ptychoramphus tenuis*
        **Mancalla diegensis*
        *Mancalla milleri
        *Mancalla emlongi
        (?) Charadriidae
Comments: Miller and DeMay (1942) cite older publications on this local fauna. *Lechusa stirtoni*, originally described from here, is a modern *Tyto alba* (Chandler 1982). Before Chandler's ongoing revision of the birds of this local fauna, *Gavia concinna*, *Podiceps parvus*, and *Mancalla californiensis* were reported as present.

Sand Draw
Formation: Keim
State: Nebraska, Brown Co.
NALMA or Correlative: Blancan
Radiometric Dates: none
General References: Skinner and Hibbard 1972
Avian References: none
Avian Taxa: Anatidae
Comments: Includes unstudied material in the AMNH Frick collections.

Sandpoint
Formation: Grand View
State: Idaho, Owyhee Co.
NALMA or Correlative: Blancan
Radiometric Dates: slightly greater than 3.0 mybp based on paleomagnetic correlation.
General References: Conrad 1980

Avian References: Murray 1970
Avian Taxa: Phalacrocoracidae
   Phalacrocorax idahensis

Santa Fe River IB
 Formation: mixed
 State: Florida, Gilchrist/Columbia Co. line
 NALMA or Correlative: Blancan
 Radiometric Dates: none
 General References: Webb 1974
 Avian References: Brodkorb 1963d
 Avian Taxa: Podicipedidae
   Podilymbus podiceps
  Phalacrocoracidae
   Phalacrocorax auritus
  Anatidae
   Aythya affinis
   Mergus merganser
  Accipitridae
   Buteo jamaicensis
  Phasianidae
   Meleagris gallopavo
  Phorusrhacidae
   *Titanis walleri
 Comments: The Santa Fe River has produced a mixed fauna of Blancan and Pleistocene fossils. It is not certain which of the "associated" taxa above are actually Blancan and which are Pleistocene.

Sawrock Canyon
 Formation: Rexroad
 State: Kansas, Seward Co.
 NALMA or Correlative: Early Blancan
 Radiometric Dates: none
 General References: Hibbard 1953, 1964

Avian References: Feduccia 1968, 1970
Avian Taxa: Podicipedidae
    Podilymbus majusculus
    Plataleidae
    Plegadis sp. (genus tentative)
    Rallidae
    Rallus sp.
    Genus indet.
    Scolopacidae
    *Tringa antiqua
Comments: Kurtén and Anderson (1980) provide a summary of the fossil vertebrates from the Rexroad Formation.

University Drive
  Formation: unreported
  State: California, Orange Co.
  NALMA or Correlative: Blancan (or Hemphillian)
  Radiometric Dates: none
  General References: none
  Avian References: Steadman 1980
  Avian Taxa: Phasianidae
    Meleagris sp.

Vallecito Creek
  Formation: Palm Spring
  State: California, San Diego Co.
  NALMA or Correlative: Blancan
  Radiometric Dates: none
  General References: Opdyke et al. 1977
  Avian References: Howard 1963, Brodkorb 1964c, Steadman 1980
  Avian Taxa: Podicipedidae
    Podiceps nigricollis
    Podiceps sp.

Anatidae
- *Anser* sp.
- ***Brantadorna** **downsi**
- *Anas* *acuta* (?)
- *Anas* *clypeata*
- ***Bucephala** **fossilis**
- *Melanitta* *perspicillata* (?)
- ***Oxyura** **bessomi**

Teratornithidae
- *Teratornis* *incredibilis*

Accipitridae
- ***Neophrontops** **vallecitoensis**
- *Aquila* *chrysaetos*
- Genus indet.

Phasianidae
- *Lophortyx* *gambeli*
- ***Meleagris** **anza**

Rallidae
- *Rallus* *limicola* (?)
- *Fulica* *americana* (?)
- ***Fulica** **hesterna**

Charadriidae
- *Charadrius* *vociferus*

Strigidae
- *Asio* sp.
- Genus and sp. indet.

Picidae
- Picinae, genus and sp. indet.

Corvidae
- Genus and sp. indet.

Fringillidae
- Genus and sp. indet.

Passeriformes, at least 4 sp.

Comments: Because this formation ranges in age from the late Hemphillian to the Irvingtonian, the avian list may include some taxa that are not strictly Blancan in age. Howard (1963) notes that the avifauna does not show significant change throughout this formation.

White Rock
    Formation: Belleville
    State: Kansas, Republic Co.
    NALMA or Correlative: late Blancan
    Radiometric Dates: none
    General References: Eshelman 1975
    Avian References: Eshelman 1975
    Avian Taxa: Anatidae
                Anas sp.
             Charadriidae
                 Pluvialis squatarola
             Passeriformes

White Bluffs
    Formation: Ringold
    State: Washington, Franklin Co.
    NALMA or Correlative: Blancan, early
    Radiometric Dates: none
    General References: Gustafson 1978
    Avian References: McKnight 1923
    Avian Taxa: Anatidae
                (?) *Aythya* sp.

Whitlock Oil Well
  Formation: Gila Group
  State: Arizona, Cochise Co.
  NALMA or Correlative: Blancan
  Radiometric Dates: none
  General References: none
  Avian References: none
  Avian Taxa: Accipitridae
              (?) <u>Accipiter</u>
  Comments: This record is from unstudied material in the AMNH Frick collections. Additional material is in the University of Arizona collections.

Literature Cited

Addicott, W.O.
- 1970. Miocene gastropods and biostratigraphy of the Kern River Area, California. United States Geological Survey Professional Paper, 642.
- 1972. Provincial middle and late molluscan stages, Temblor Range, California. In Pacific Coast Miocene Biostratigraphic Symposium: Proceedings of the Society of Economic Paleontology and Mineralogy, Pacific Section, March 1972, Bakersfield, California, pages 1-26.

Alvarez, R. and S.L. Olson
- 1978. A new merganser from the Miocene of Virginia (Aves: Anatidae). Proceedings of the Biological Society of Washington, 91:522-532.

Arellano, A. and E. Azcon
- 1949. Pre-Equus horses from Goleta (Morelia) Michoacan, Mexico. [Abstract]. Bulletin of the Geological Society of America, 60:1871.

Armentrout, J.M.
- 1981. Correlation and ages of Cenozoic chronostratigraphic units in Oregon and Washington. Geological Society of America, Special Paper, 184:137-148.

Barbour, E.
- 1913. Mammalian fossils from Devils' Gulch. Nebraska Geological Survey, 4:77-190.

Barghoorn, S.
- 1981. Magnetic-polarity stratigraphy of the Miocene type Tesuque Formation, Santa Fe Group, in the Española Valley, New Mexico. Bulletin of the Geological Society of America, 92:1027-1041.

Barnes, L.G.
- 1972. Miocene Desmatophocinae (Mammalia: Carnivora) from California. *University of California Publication in Geological Science*, 89:1-68.
- 1973. *Praekogia cedrosensis*, a new genus and species of fossil pygmy sperm whale from Isla Cedros, Baja California, Mexico. *Contributions in Science, Los Angeles County Museum of Natural History*, 247:1-20.
- 1976. Outline of eastern North Pacific fossil cetacean assemblages. *Systematic Zoology*, 25:321-343.
- 1978. A review of *Lophocetus* and *Liolithax* and their relationships to the delphinoid family Kentriodontidae (Cetacea: Odontoceti). *Bulletin of the Los Angeles County Museum of Natural History*, 28:1-35.
- 1979. Fossil enaliarctine pinnipeds (Mammalia: Otariidae) from Pyramid Hill, Kern County, California. *Contributions in Science, Los Angeles County Museum of Natural History*, 318:1-41.
- 1984. Fossil odontocetes (Mammalia: Cetacea) from the Almejas Formation, Isla Cedros, Mexico. *PaleoBios*, 42:1-46.

Barnes, L.G., H. Howard, J.H. Hutchison, and B.J. Welton
- 1981. The vertebrate fossils of the marine Cenozoic San Mateo Formation at Oceanside, California. In P.L. Abbott and S. O'Dunn. *Geologic Investigations of the San Diego costal plain*. *San Diego Association of Geologists*, San Diego, California. pages 53-70.

Barnes, L.G., C.P. Domning, H. Howard, R.W. Huddleston, and C.A. Repenning

In prep. Correlation and characterization of Late Miocene (Clarendonian correlative) Marine Vertebrate Assemblages in California.

Baum, G.R. and W.H. Wheeler
- 1977. Cetaceans from the St. Marys and Yorktown Formations, Surry County, Virginia. Journal of Paleontology, 51:492-504.

Becker, J.J.
- 1980. The Star Valley local fauna (Early Hemphillian), southwestern Idaho. 71 pages. Master's Thesis, Department of Biology, Idaho State University, Pocatello, Idaho.
- 1985a. Fossil herons (Aves: Ardeidae) of the late Miocene and early Pliocene of Florida. Journal of Vertebrate Paleontology, 5:24-31.
- 1985b. Pandion lovensis, a new species of Osprey from the late Miocene of Florida. Proceedings of the Biological Society of Washington, 98:314-320.
- 1985c. The fossil birds of the late Miocene and early Pliocene of Florida. 245 pages. Ph.D. Dissertation, University of Florida, Gainesville, Florida.
- 1986a. An early heron (Aves, Ardeidae, Ardea) from the Middle Miocene of Nebraska. Journal of Paleontology, 60:968-970.
- 1986b. Re-identification of "Phalacrocorax subvolans" Brodkorb as the earliest record of Anhingidae. Auk, 103:804-808.
- 1986c. A fossil Motmot (Aves: Momotidae) from the late Miocene of Florida. Condor, 88:478-482.
- 1986d. A new vulture (Vulturidae, Pliogyps) from late Miocene of Florida. Proceedings of the Biological Society of Washington, 99:502-508.

1986e. Birds of the Pliocene (Blancan) Oreana local fauna, Owyhee County, Idaho. Great Basin Naturalist, 46:763-768.

1987a. Systematic revision of "Falco" ramenta Wetmore and the Neogene evolution of the Falconidae. Auk, 105:270-276.

1987b. Additional material of Anhinga grandis Martin and Mengel (Aves: Anhingidae) from the late Miocene of Florida. Proceeding of the Biological Society of Washington 100: 358-363.

Bennett, D.K.
1984. Cenozoic rocks and faunas of north-central Kansas. 128 pages. Ph.D. Dissertation, University of Kansas, Lawrence, Kansas.

Berggren, W.A. and J.A. Van Couvering
1974. The Late Neogene: Biostratigraphy, geochronology, and paleoclimatology of the last 15 million years in marine and continental sequences. Palaeogeography, Palaeoclimatology, and Palaeoecology, 16:1-216.

Berggren, W.A., D.V Kent, J.J. Flynn, and J.A. Van Couvering
1985. Cenozoic geochronology. Bulletin of the Geological Society of America, 96:1407-1418.

Berta, A. and G.S. Morgan
1985. A new sea otter (Carnivora: Mustelidae) from the late Miocene and early Pliocene of North America. Journal of Paleontology, 59:809-819.

Bickart, K.J.
1981. A new Thick-knee, Burhinus, from the Miocene of Nebraska, with comments on the habitat requirements of the Burhinidae (Aves: Charadriiformes). Journal of Vertebrate Paleontology, 1:273-277.

1986. The birds of the late Miocene -- early Pliocene Big Sandy Formation, Mohave County, Arizona. 126 pages. Master's Thesis. University of Michigan, Ann Arbor, Michigan.

Bjork, P.
1970. The carnivora of the Hagerman local fauna (Late Pliocene) of southwestern Idaho. Transactions of the American Philosophical Society, new series, 60(7):1-54.

Blackwelder, B.W.
1981. Late Cenozoic stages and molluscan zones in the U.S. Atlantic Coastal Plain. [Paleontological Society Memoir 12], Journal of Paleontology, 55(5, Supl.):1-35.

Blackwelder, B.W. and L.W. Ward
1976. Stratigraphy of the Chesapeake Group of Maryland and Virginia. In Geological Society of America, Guidebook for Fieldtrip 7b, Arlington, Virginia. 55 pages.

Blow, W.H.
1969. Late Middle Eocene to Recent planktonic foraminiferal biostratigraphy. In: P. Brönnimann and H.H. Renz. Proceedings of the International Conference on Planktonic Microfossils, 1st., Geneva 1967, 1. E.J. Brill, Leiden, pages 199-421.

Breyer, J.
1981. The Kimballian Land-Mammal Age: Mene, Mene, Tekel, Upharsin (Dan. 5:25). Journal of Paleontology, 55:1207-1216.

Brodkorb, P.
1952. The types of Lambrecht's fossil bird genera. Condor, 54:174-175.

1953a. A Pliocene grebe from Florida. Annals and Magazine of Natural History, series 12, 6:953-954.

1953b. A Pliocene flamingo from Florida. Chicago Academy of Sciences, Natural History Miscellany, 124:1-3.

1953c. A Pliocene gull from Florida. Wilson Bulletin, 65:94-98.

1953d. A review of the Pliocene loons. Condor, 55:211-214.

1954. A chachalaca from the Miocene of Florida. Wilson Bulletin, 66:180-183.

1955. The avifauna of the Bone Valley Formation. Florida Geological Survey, Report of Investigations, 14:1-57.

1956. Two new birds from the Miocene of Florida. Condor, 58:367-370.

1958a. Birds from the middle Pliocene of McKay, Oregon. Condor, 60:252-255.

1958b. Fossil birds from Idaho. Wilson Bulletin, 70:237-242.

1961. Birds from the Pliocene of Juntura, Oregon. Quarterly Journal of the Florida Academy of Sciences, 24:169-184.

1962. A teal from the lower Pliocene of Kansas. Quarterly Journal of the Florida Academy of Sciences, 25:157-160.

1963a. Fossil birds from the Alachua clays of Florida. Florida Geological Survey, Special Publication 2, Paper, 4:1-17.

1963b. Miocene birds from the Hawthorne formation. Quarterly Journal of the Florida Academy of Sciences, 26:159-167.

1963c. Catalogue of Fossil Birds. Part 1 (Archaeopterygiformes through Ardeiformes).

Bulletin of the Florida State Museum, Biological Sciences, 7:180-293.

1963d. A giant flightless bird from the Pleistocene of Florida. Auk, 80:111-115.

1964a. A Pliocene teal from South Dakota. Quarterly Journal of the Florida Academy of Sciences, 27:55

1964b. Catalogue of Fossil Birds. Part 2 (Anseriformes through Galliformes). Bulletin of the Florida State Museum, Biological Sciences, 8:195-335.

1964c. Notes on fossil turkeys. Quarterly Journal of the Florida Academy of Sciences, 27:223-229.

1967. Catalogue of Fossil Birds. Part 3 (Ralliformes, Icthyornithiformes, Charadriiformes). Bulletin of the Florida State Museum, Biological Sciences, 11:99-220.

1969. An ancestral Mourning Dove from Rexroad, Kansas. Quarterly Journal of the Florida Academy of Sciences, 31:173-176.

1970. New discoveries of Pliocene birds in Florida. [Abstract]. Proceedings of the XVth International Ornithological Congress, page 74. Leiden, E.J. Brill.

1971a. Catalogue of Fossil Birds. Part 4 (Columbiformes through Piciformes). Bulletin of the Florida State Museum, Biological Sciences, 15:163-266.

1971b. The paleospecies of Woodpeckers. Quarterly Journal of the Florida Academy of Sciences, 33:132-136.

1972. Neogene fossil jays from the Great Plains. Condor, 74:347-349.

1978. Catalogue of Fossil Birds. Part 5 (Passeriformes). *Bulletin of the Florida State Museum, Biological Sciences*, 23:139-228.

Burt, W.H.
1929. A new goose (*Branta*) from the lower Pliocene of Nevada. *University of California Publication, Bulletin of the Department of Geological Sciences*, 18:221-224.

Campbell, K.E., Jr.
1976. An early Pleistocene avifauna from Haile XVA, Florida. *Wilson Bulletin*, 88:345-347.

Chandler, R.M.
1982. A reevaluation of the Pliocene owl *Lechusa stirtoni* Miller. *Auk*, 99:580-581.

Coats, R.R.
1985. The Big Island Formation, a Miocene formation in northern Elko County, Nevada, and adjacent Idaho, including a consideration of its composition and petrographic character. *United States Geological Survey Bulletin*, 1605-A:A7-A13.

Collins, C.
1964. Fossil ibises from the Rexroad fauna of the upper Pliocene of Kansas. *Wilson Bulletin*, 76:43-49.

Compton, L.
1934. Fossil birds from the Pliocene and Pleistocene of Texas. *Condor*, 36:40-41.
1935a. Two avian fossils from the lower Pliocene of South Dakota. *American Journal of Science*, 30:343-348.
1935b. An anserine fossil from the Pliocene of western Nebraska. *Condor*, 37:43-44.

Conrad, G.S.
1980. The biostratigraphy and mammalian paleontology of the Glenns Ferry Formation from Hammett to Oreana, Idaho. 351 pages. Ph.D. Dissertation, Idaho State University, Pocatello, Idaho.

Cook, H.J. and M. Cook
1933. Faunal lists of the Tertiary Vertebrata of Nebraska and adjacent areas. Nebraska Geological Survey Paper, 5. 58 pages.

Cope, E.D.
1870. Synopsis of the extinct Batrachia, Reptilia, and Aves of North America. Transactions of the American Philosophical Society, new series, 14: i-viii, 1-252, pls. i-xiv. [Pages 1 - 104 appeared in 1869, pp. 105 - 234 in April 1870, and the remaining pages in December 1870; imprint is 1871].

1874. Notes on the Santa Fe marls and some of the contained vertebrate fossils. Proceedings of the Academy of Natural Sciences, Philadelphia, 26:147-152.

1894. On Cyphornis, an extinct genus of birds. Journal of the Academy of Natural Sciences, Philadelphia, series 2, 9:449-452.

Cracraft, J.
1971. The humerus of the early Miocene cracid, Boreortalis laesslei Brodkorb. Wilson Bulletin, 83:200-201.

Cracraft, J. and J. Morony
1969. A new Pliocene woodpecker, with comments on the fossil picidae. American Museum Novitates, 2400:1-8.

Dalquest, W.W.
1983. Mammals of the Coffee Ranch local fauna Hemphillian of Texas. *Texas Memorial Museum, Pearce-Sellards Series*, 38:1-41.

Dalquest, W.W. and T.J. Donovan
1973. A new three-toed horse (*Nannipus*) from the late Pliocene of Scurry County, Texas. *Journal of Paleontology*, 47:34-35.

Domning, D.P., C.E. Ray, and M.C. McKenna
1986. Two new Oligocene Desmostylians and a discussion of Tethytherian systematics. *Smithsonian Contributions to Paleobiology*, 59:1-56.

Downs, T.
1956. The Mascall Fauna from the Miocene of Oregon. *University of California Publication, Bulletin of the Department of Geological Sciences*, 31:199-354.

Ekren, E.B., D.H. McIntyre, and E.H. Bennett
1978. Preliminary geological map of the west half of Owyhee County, Idaho. *United States Geological Survey Open File Report*, 78-341.

Elias, M.K.
1942. Tertiary prairie grasses and other herbs from the High Plains. *Geological Society of America, Special Papers*, 41. 176 pages.

Eshelman, R.E.
1975. Geology and paleontology of the early Pleistocene (late Blancan) White Rock Fauna from north-central Kansas. *University of Michigan, Museum of Paleontology, Papers on Paleontology*, 13:1-60.

Evernden, J.F., D.E. Savage, G.H. Curtis, and G.T. James
    1964. Potassium-Argon dates and the Cenozoic mammalian chronology of North America. American Journal of Science, 262:145-198.

Feduccia, J.A.
    1967a. A new swallow from the Fox Canyon local fauna (upper Pliocene) of Kansas. Condor, 69:526-527.
    1967b. Ciconia maltha and Grus americana from the upper Pliocene of Idaho. Wilson Bulletin, 79:316-318.
    1968. The Pliocene Rails of North America. Auk, 85:441-453.
    1970. A new shorebird from the upper Pliocene. Journal of the Graduate Research Center, Southern Methodist University Press, 38:58-60.
    1974. Another Old World Vulture from the New World. Wilson Bulletin, 86:251-255.
    1975. Professor Hibbard's Fossil Birds. University of Michigan, Museum of Paleontology, Papers on Paleontology, 12:67-70.

Feduccia, J.A. and N.L. Ford
    1970. Some birds of prey from the upper Pliocene of Kansas. Auk, 87:795-797.

Feduccia, J.A. and R.L. Wilson
    1967. Avian fossils from the lower Pliocene of Kansas. Occasional Papers of the Museum of Zoology, University of Michigan, 655. 2 pages.

Fife, D.
    1974. Geology of the south half of the El Toro Quadrangle, Orange County, California. California Division of Mines and Geology, Special Report, 110:1-27.

Ford, N.L.
- 1966. Fossil owls from the Rexroad fauna of the upper Pliocene of Kansas. Condor, 68:472-475.
- 1967. A systematic study of the owls based on comparative osteology. 128 pp. Ph.D. Dissertation, University of Michigan, Ann Arbor, Michigan.

Ford, N.L. and B.G. Murray, Jr.
- 1967. Fossil owls from the Hagerman local fauna (upper Pliocene) of Idaho. Auk, 84:115-117.

Galbreath, E.C.
- 1953. A contribution to the Tertiary geology and paleontology of northeastern Colorado. University of Kansas Paleontology, Vertebrata. No. 4.
- 1964. A corvid from the Miocene of Colorado. Transactions of the Illinois State Academy of Sciences, 57:282.

Galusha, T.
- 1966. The Zia Sand Formation, new early to medial Miocene beds in New Mexico. American Museum Novitates, 2271. 12 pages.
- 1975. Stratigraphy of the Box Butte Formation Nebraska. Bulletin of the American Museum of Natural History, 156:1-68.

Galusha, T. and J.C. Blick
- 1971. Stratigraphy of the Santa Fe Group, New Mexico. Bulletin of the American Museum of Natural History, 144:1-127.

Galusha, T., N.M. Johnson, E.H. Lindsay, N.D. Opdyke, and R.H. Tedford
- 1984. Biostratigraphy and magnetostratigraphy, late Pliocene rocks, 111 Ranch, Arizona. Bulletin of the Geological Society of America, 95:714-722.

Gawne, C.E.
  1981. Sedimentology and stratigraphy of the Miocene Zia Sand of New Mexico: Summary. Bulletin of the Geological Society of America, Part 1. 92:999-1007.

Gazin, C.L.
  1942. The late Cenozoic vertebrate faunas from San Pedro Valley, Arizona. Proceedings of the United States National Museum, 92:475-518.

Gazin, C.L. and R.L. Collins
  1950. Remains of land mammals from the Miocene of the Chesapeake Bay Region. Smithsonian Miscellaneous Collection, 116(2):1-21.

Gregory, J.
  1942. Pliocene vertebrates from Big Spring Canyon, South Dakota. University of California Publication, Bulletin of the Department of Geological Sciences, 26:307-446.

Gustafson, E.P.
  1978. The vertebrate faunas of the Pliocene Ringold Formation, south-central Washington. Bulletin of the Museum of Natural History, University of Oregon, 23:1-62.

Harksen, J. and J.R. Macdonald
  1967. Miocene Batesland Formation named in southwestern South Dakota. South Dakota Geological Survey, Report of Investigations, No. 96.

Harrison, J.A.
  1981. A review of the extinct Wolverine, Pleisogulo (Carnivora: Mustelidae), from North America. Smithsonian Contributions of Paleobiology, 46:1-27.

Harrison, C.J.O. and C. Walker
    1976. A review of the Bony-toothed birds (Odontopterygiformes): with descriptions of some new species. *Tertiary Research, Special Paper*, No. 2. 72 pages.

Hesse, C.J.
    1935. A vertebrate fauna from the type locality of the Ogallala Formation. *University of Kansas Science Bulletin*, 22:79-101.

Hibbard, C.W.
    1950. Mammals of the Rexroad Formation Fox Canyon, Kansas. *Contribution of the Museum of Paleontology, University of Michigan*, 8:113-192.
    1953. The Saw Rock Canyon fauna and its stratigraphic significance. *Papers of the Michigan Academy of Science, Arts, and Letters*, 38:387-411.
    1964. A contribution to the Saw Rock local fauna of Kansas. *Papers of the Michigan Academy of Science, Arts, and Letters*, 49:115-127.
    1970. Pleistocene mammalian faunas from the Great Plains and central lowland provinces of the United States. *In* W. Dort, Jr. and J.K. Jones, Jr. *Pleistocene and Recent Environments of the Central Great Plains*. Department of Geology, University of Kansas Special Publication, 3:395-433.

Hirschfeld, S. and S.D. Webb
    1968. Plio-Pleistocene megalonychid sloths of North America. *Bulletin of the Florida State Museum, Biological Sciences*, 55:213-296.

Holman, J.
    1961. Osteology of living and fossil New World Quails (Aves: Galliformes). *Bulletin of the Florida State Museum, Biological Sciences*, 6:131-233.

Howard, H.
- 1944. A Miocene hawk from California. *Condor*, 46:236-237.
- 1949. New avian records for the Pliocene of California. *Carnegie Institution of Washington, Publication*, 584(6):177-199.
- 1957a. A gigantic "toothed" marine bird from the Miocene of California. *Santa Barbara Museum of Natural History, Bulletin of the Department of Geology*, 1:1-23.
- 1957b. A new species of passerine bird from the Miocene of California. *Contributions in Science, Los Angeles County Museum of Natural History*, 9:1-16.
- 1958. Miocene sulids of southern California. *Contributions in Science, Los Angeles County Museum of Natural History*, 25:1-16.
- 1962. A new Miocene locality for *Puffinus diatomicus* and *Sula willetti*. *Condor*, 64:512-513.
- 1963. Fossil birds from the Anza-Borrego Desert. *Contributions in Science, Los Angeles County Museum of Natural History*, 73:1-33.
- 1965. A new species of cormorant from the Pliocene of Mexico. *Bulletin of the Southern California Academy of Sciences*, 64:50-55.
- 1966a. Pliocene birds from Chihuahua, Mexico. *Contributions in Science, Los Angeles County Museum of Natural History*, 94:1-12.
- 1966b. A possible ancestor of the Lucas Auk (Family Mancallidae) from the Tertiary of Orange County, California. *Contributions in Science, Los Angeles County Museum of Natural History*, 101:1-8.
- 1966c. Additional avian records from the Miocene of Sharkstooth Hill, California. *Contributions in*

Science, Los Angeles County Museum of Natural History, 114:1-11.

1968. Tertiary birds from Laguna Hill, Orange County, California. Contributions in Science, Los Angeles County Museum of Natural History, 142:1-21.

1969. A new avian fossil from Kern Co., California. Condor, 71:68-69.

1970. A review of the extinct avian genus, Mancalla. Contributions in Science, Los Angeles County Museum of Natural History, 203:1-12.

1971. Pliocene avian remains from Baja California. Contributions in Science, Los Angeles County Museum of Natural History, 217:1-17.

1972. Type specimens of avian fossils in the collections of the Natural History Museum of Los Angeles County. Contributions in Science, Los Angeles County Museum of Natural History, 228:1-27.

1976. A new species of flightless auk from the Miocene of California (Alcidae: Mancallinae). Smithsonian Contributions in Paleobiology, 27:141-146.

1978. Late Miocene marine birds from Orange Co., California. Contributions in Science, Los Angeles County Museum of Natural History, 290:1-28.

1981. A new species of Murre, Genus Uria from the late Miocene of California (Aves: Alcidae). Bulletin of the Southern California Academy of Sciences, 80:1-12.

1982. Fossil birds from Tertiary marine beds at Oceanside, San Diego County, California with descriptions of two new species of the genera Uria and Cepphus (Aves: Alcidae).

   Contributions in Science, Los Angeles County Museum of Natural History, 341:1-15.
1984. Additional records from the Miocene of Kern County, California with the description of a new species of Fulmar (Aves: Procellariidae). Bulletin of the Southern California Academy of Sciences, 83:84-89

Howard, H. and J.A. White
1962. A second record of Osteodontornis, Miocene "toothed" bird. Contributions in Science, Los Angeles County Museum of Natural History, 52:1-12.

Hunt, R.M., Jr.
1972. Miocene amphicyonids (Mammalia: Carnivora) from Agate Springs quarries, Sioux County, Nebraska. American Museum Novitates, 2506. 39 pages.
1981. Geology and vertebrate paleontology of the Agate Fossil Beds National Monument and surrounding region, Sioux county, Nebraska (1972-1978). National Geographic Society Research Reports, 13:263-285.

Hunter, H.E. and P.F. Huddlestun
1982. The biostratigraphy of the Torreya Formation of Florida. Pages 211-223. In T.M. Scott and S.B. Upchurch, Miocene of the southeastern United States. Florida Bureau of Geology, Special Publication No. 25.

Izett, G.A. and C.W. Naesler
1981. Fisson-track ages of airfall tuffs in Miocene sedimentary rocks of the Española Basin, Santa Fe County, New Mexico. United States Geological Survey Open-File Report, 81-161, 6 pages.

Izett, G.A., R.W. Wilcox, and G.A. Borchardt
    1972. Correlation of a volcanic ash near Mount Blanco, Texas with the Guaje pumice bed of the Jemex Mountains, New Mexico. Quaternary Research, 2:554-578.

Jacobs, L. and E.H. Lindsay
    1981. Prosigmodon oroscoi, a new sigmodont rodent from the Late Tertiary of Mexico. Journal of Paleontology, 55:425-430.

Jahns, R.H.
    1940. Stratigraphy of the easternmost Ventura Basin California, with a description of a new lower Miocene mammalian fauna from the Tick Canyon Formation. Carnegie Institution of Washington, Publication, 514:145-194.

Johnsgard, P.A.
    1979. Order Anseriformes. In E. Mayr and G. W. Cottrell, Check-list of Birds of the World, vol. 1, Second edition, pages 425-506. Cambridge, Mass., Museum of Comparative Zoology.

Johnstone, C.S. and D.E. Savage
    1955. A survey of various Cenozoic vertebrate faunas of the panhandle of Texas. Part. 1. Introduction, description of localities, preliminary faunal lists. University of California Publication, Bulletin of the Department of Geological Sciences, 31:27-50.

Kellogg, R.
    1932. A Miocene long-beaked porpoise from California. Smithsonian Miscellaneous Collections, 87(2):1-11.
    1934. A new cetathere from the Modelo Formation at Los Angeles, California. Carnegie Institution of Washington, Publication, 447:83-104.

Kimmel, P.G.
 1979. Stratigraphy and paleoenvironments of the Miocene Chalk Hills Formation in the western Snake River Plain, Idaho. 340 pages. Ph.D. Dissertation, University of Michigan, Ann Arbor, Michigan.

Kurtén, B. and E. Anderson
 1980. Pleistocene mammals of North America. 442 pages. New York: Columbia University Press.

Lambrecht, K.
 1933. Handbuch der Palaeornithologie. 1024 pages. Berlin, Gebrüder Borntraeger.

Lance, J.
 1950. Paleontologia y estratigrafia del Plioceno de Yepomera, estado de Chihuahua. Pt. 1: Equidas, excepto Neohipparion. Universidad Nacional Autonoma Mexico, Instituto Geologia, Bul. 54:viii+81 pp.

Leidy, J. and F. Lucas
 1896. Fossil vertebrates from the Alachua clays. Transactions of the Wagner Free Institute Science, 4:1-61.

Lindsay, E.H.
 1972. Small mammals from the Barstow Formation, California. University of California Publication, Bulletin of the Department of Geological Sciences, 93:1-104.

Lindsay, E.H., N.M. Johnson, and N.D. Opdyke
 1975. Preliminary correlation of North American Land Mammal Ages and geomagnetic chronology. University of Michigan Papers on Paleontology, 12:111-119.

Lucas, F.A.
 1901a. A fossil, flightless Auk. Science, new series, 13(324):428.

1901b. A flightless Auk, Mancalla californiensis, from the Miocene of California. *Proceedings of the United States National Museum*, 24(1245):133.

Macdonald, J.R.
1948. The Pliocene carnivores of the Black Hawk Ranch Fauna. *University of California Publication, Bulletin of the Department of Geological Sciences*, 28:53-80.

MacFadden, B.J.
1982. New species of primitive three-toed browsing horse from the Miocene phosphate mining district of Central Florida. *Florida Scientist*, 45:117-125.

MacFadden, B.J., N.M. Johnson, and N.D. Opdyke
1979. Magnetic polarity stratigraphy of the Mio-Pliocene mammal-bearing Big Sandy Formation of western Arizona. *Earth and Planetary Science Letters*, 44:349-364.

MacFadden, B.J. and S.D. Webb.
1982. The succession of Miocene (Arikareean through Hemphillian) terrestrial mammalian localities and faunas in Florida. Pages 186-199. In T.M. Scott and S.B. Upchurch, *Miocene of the southeastern United States*. Florida Bureau of Geology, Special Publication No. 25.

Marsh, O.C.
1870. Notice of some fossil birds, from the Cretaceous and Tertiary Formations of the United States. *American Journal of Science, second series*, 49(146):205-217.

1871. Notice of some new fossil mammals and birds, from the Tertiary formations of the West. *American Journal of Science, series 3*, 2:120-127.

1893. Description of Miocene mammalia. American Journal of Science, 46:407-412.

Martin, H.T.
1928. Two new carnivores from the Pliocene of Kansas. Journal of Mammalogy, 9:233-236.

Martin, J.E.
1978. A new and unusual Shrew (Soricidae) from the Miocene of Colorado and South Dakota. Journal of Paleontology, 52:636-641.
1985. Geological and Paleontological road log from Rapid City, through the Oligocene White River Badlands and Miocene deposits, to Pine Ridge, South Dakota. pp. 13-119. In J.E. Martin, Fossiliferous Cenozoic deposits of western South Dakota and northwestern Nebraska. Dakoterra 2(2). 367 pages. [Guidebook for the 45th annual meeting of the Society of Vertebrate Paleontology].

Martin, L.D.
1975. A new species of Spizaetus from the Pliocene of Nebraska. Wilson Bulletin, 87:413-416.

Martin, L.D. and R. Mengel
1975. A new species of Anhinga (Anhingidae) from the upper Pliocene of Nebraska. Auk, 92:137-140.
1980. A new goose from the late Pliocene of Nebraska with notes on variability and proportions in some recent geese. Contributions in Science, Los Angeles County Museum of Natural History, 330:75-85.
1984. A new cuckoo and a chachalaca from the early Miocene of Colorado. Special Publication of the Carnegie Museum of Natural History, 9:171-177.

Martin, L.D. and J. Tate
- 1970. A new Turkey from the Pliocene of Nebraska. *Wilson Bulletin*, 82:214-218.

Matthew, W.D.
- 1924. Third contribution to the Snake Creek fauna. *Bulletin of the American Museum of Natural History*, 50:59-210.

Matthew, W.D. and H.J. Cook
- 1909. A Pliocene fauna from western Nebraska. *Bulletin of the American Museum of Natural History*, 26:361-414.

Mawby, J.
- 1965. Pliocene vertebrates and stratigraphy in Stewart and Ione Valleys, Nevada. 208 pages. Ph.D. Dissertation, University of California, Berkeley.
- 1968a. *Megabelodon minor* (Mammalia, Proboscidea), a new species of mastodont from the Esmeralda Formation of Nevada. *PaleoBios*, 4:1-10.
- 1968b. *Megahippus* and *Hypohippus* (Perissodactyla, Mammalia) from the Esmeralda Formation of Nevada. *PaleoBios*, 7:1-13.

May, S.R.
- 1981. *Repomys* (Mammalia: Rodentia gen. nov.) from the late Neogene of California and Nevada. *Journal of Vertebrate Paleontology*, 1:219-230.

Merriam, J.C.
- 1896. Note on two Tertiary faunas from the rocks of the southern coast of Vancouver Island. *University of California Publication, Bulletin of the Department of Geology*, 2:101-108.
- 1915. Tertiary vertebrate faunas of the North Colinga region of California. *Transactions of the American Philosophical Society*, 22:191-124.

   1919. Tertiary mammalian faunas of the Mohave Desert. University of California Publication, Bulletin of the Department of Geological Sciences, 11:462.

McKnight, E.T.
   1923. The White Bluffs Formation of the Columbia. Thesis, University of Washington, Seattle, Washington.

Miller, A.H.
   1944. An avifauna from the lower Miocene of South Dakota. University of California Publication, Bulletin of the Department of Geological Sciences, 27(4):85-100.
   1948. The Whistling Swan in the upper Pliocene of Idaho. Condor, 50:132.

Miller, A.H. and R.I. Bowman
   1956a. Fossil birds of the late Pliocene of Cita Canyon, Texas. Wilson Bulletin, 68:38-46.
   1956b. A fossil magpie from the Pleistocene of Texas. Condor, 58:164-165.

Miller, A.H. and L.V. Comptom
   1939. Two fossil birds from the lower Miocene of South Dakota. Condor, 41:153-156.

Miller, A.H. and C. Sibley
   1941. A Miocene gull from Nebraska. Auk, 58:563-666.
   1942. A new species of crane from the Pliocene of California. Condor, 44:126-127.

Miller, L.
   1925. Avian remains from the Miocene of Lompoc. Carnegie Institution of Washington, Publication, 349:107-117.
   1929. A new cormorant from the Miocene of California. Condor, 31:167-172.
   1930. A fossil goose from the Ricardo Pliocene. Condor, 32:208-209.

1931. Bird remains from the Kern River Pliocene of California. Condor, 33:70-72.

1935. New bird horizons in California. Publication of the University of California, Los Angeles, Biology, 1:783-80.

1937. An extinct puffin from the Pliocene of San Diego, California. Transactions of the San Diego Society of Natural History, 8:375-378.

1944a. A Pliocene flamingo from Mexico. Wilson Bulletin, 56:77-82.

1944b. Some Pliocene Birds from Oregon and Idaho. Condor, 46:25-32.

1950. A Miocene flamingo from California. Condor, 52:69-73.

1951. A Miocene petrel from California. Condor, 53:78-80.

1952. The avifauna of the Barstow Miocene of California. Condor, 54:296-301.

1956. A collection of bird remains from the Pliocene of San Diego, California. Proceedings of the California Academy of Sciences, 28:615-621.

1960. On the history of the Cathartidae in North America. Novitates Colombianas, 1:232-235.

1961. Birds from the Miocene of Sharkstooth Hill, California. Condor, 63:399-402.

1962. A new albatross from the Miocene of California. Condor, 64:471-472.

1966. An addition to the bird fauna of the Barstow Miocene. Condor, 68:397.

Miller, L. and R.I. Bowman

1958. Further bird remains from the San Diego Pliocene. Contributions in Science, Los Angeles County Museum of Natural History, 20:1-15.

Miller, L. and I. DeMay
- 1942. Fossil birds of California, an avifauna and bibliography with annotations. *University of California Publication in Zoology*, 47:47-142.

Miller, L. and H. Howard
- 1949. The flightless Pliocene bird, *Mancalla*. *Carnegie Institution of Washington, Publication*, 584:201-228.

Miller, L. and C.S. Johnstone
- 1937. A Pliocene record of *Parapavo* from Texas. *Condor*, 39:229.

Mitchell, E.D.
- 1965. History of research at Sharkstooth Hill. *Special Publication of the Kern County Historical Society*, pages i-vi, 1-45.
- 1966. Faunal succession of extinct North Pacific marine mammals. *Norsk Hvalfangst-Tidende*, 1966(3):47-60.

Mitchell, E.D. and R.H. Tedford
- 1973. The Enaliarctinae: a new group of extinct aquatic Carnivora and a consideration of the Otariidae. *Bulletin of the American Museum of Natural History*, 151:201-284.

Mosley, C. and J.A. Feduccia
- 1975. Upper Pliocene herons and Ibises from North America. *University of Michigan, Museum of Paleontology, Papers in Paleontology*, 12:71-74.

Murray, B.G., Jr.
- 1967. Grebes from the Late Pliocene of North America. *Condor*, 69:277-288.
- 1970. A redescription of two Pliocene cormorants. *Condor*, 72:293-298.

Nomland, J.
- 1916. Relationship of the invertebrate to the vertebrate faunal zones of the Jacalitos and

Etchegoin formations in the North Colinga region, California. *University of California Publication, Bulletin of the Department of Geological Sciences*, 9:77-88.

Newcomb, R.C.
1971. Relationship of the Ellensburg Formation to extensions of the Dalles Formation in the area of Arlington and Shutler Flat, north Central Oregon. *OreBin*, 33(7):133-142.

Olsen, S.J.
1964. Vertebrate correlation and Miocene stratigraphy of north Florida fossil localities. *Journal of Paleontology*, 38:600-604.

Olson, S.L.
1976. A Jacana from the Pliocene of Florida (Aves: Jacanidae). *Proceedings of the Biological Society of Washington*, 89:259-264.

1977a. A great Auk, *Pinquinis* [sic], from the Pliocene of North Carolina (Aves: Alcidae). *Proceedings of the Biological Society of Washington*, 90:690-697.

1977b. A synopsis of the fossil Rallidae. In S.D. Ripley, *Rails of the World. A monograph of the family Rallidae*. pages 339-373. Boston: D. R. Godine.

1981a. The generic allocation of *Ibis pagana* Milne Edwards, with a review of fossil Ibises (Aves: Threskiornithidae). *Journal of Vertebrate Paleontology*, 1:165-170.

1981b. A third species of *Mancalla* from the late Pliocene San Diego Formation of California (Aves: Alcidae). *Journal of Vertebrate Paleontology*, 1:97-99.

1984. A brief synopsis of the fossil birds from the Pamunkey River and other Tertiary marine deposits in Virginia. In L.W. Ward and K. Krafft, Stratigraphy and paleontology of the outcropping Tertiary beds in the Pamunkey River region, central Virginia Coastal Plain. Guidebook for Atlantic Coastal Plain Geological Association 1984 fieldtrip: Atlantic Coastal Plain Geological Association.

1985a. A new genus of tropicbird (Pelecaniformes: Phaethontidae) from the middle Miocene Calvert Formation of Maryland. Proceedings of the Biological Society of Washington, 98:851-855.

1985b. The fossil record of birds. Pages 76-252. In D.S. Farner and J. King. Avian Biology, vol. 8. Orlando, Academic Press.

Olson, S.L. and J. Farrand

1974. Rhegminornis restudied: a tiny Miocene turkey. Wilson Bulletin, 86:114-120.

Olson, S.L. and A. Feduccia

1980. Relationships and evolution of Flamingos (Aves: Phoenicopteridae). Smithsonian Contributions to Zoology, 316. 73 pages.

Olson, S.L. and D.D. Gillette

1978. Catalogue of type specimens of fossil vertebrates Academy of Natural Sciences, Philadelphia Part III: Birds. Proceedings of the Academy of Natural Sciences of Philadelphia, 129:99-100.

Olson, S.L. and D.W. Steadman

1978. The fossil record of the Glareolidae and Haematopodidae (Aves: Charadriiformes). Proceedings of the Biological Society of Washington, 91:972-981.

Opdyke, N.D., E.H. Lindsay, N.M. Johnson, and T. Downs

1977. The paleomagnetism and magnetic polarity stratigraphy of the mammal-bearing section of the Anza Borrego State Park, California. *Quaternary Research*, 7:316-329.

Orr, R.T.
    1940. Another record of *Puffinus diatomicus*. *Auk*, 57:105.

Patton, T.
    1967. Oligocene and Miocene vertebrates from central Florida. *Southeastern Geological Society, 13th Field Trip Guidebook*, pages 3-10.

Ray, C.E.
    1976. Geography of Phocid evolution. *Systematic Zoology*, 25:391-406.
    1983. [Editor]. Geology and paleontology of the Lee Creek Mine, North Carolina, I. *Smithsonian Contributions to Paleobiology*, 53. 529 pp.

Reed, L. and O. Longnecker, Jr.
    1932. The geology of Hemphill County, Texas. *University of Texas Bulletin*, 3231, 98 pages.

Repenning, C.A.
    1962. The giant ground squirrel *Paenemarmota*. *Journal of Paleontology*, 36:540-556.
    1976. *Enhydra* and *Enhydriodon* from the Pacific Coast of North America. *Journal of Research, United States Geological Survey*, 4:305-315.

Repenning, C.A. and R.H. Tedford
    1977. Otarioid seals of the Neogene. *United States Geological Survey Professional Paper*, 992.

Rich, P.V.
    1980. "New World Vultures" with Old World affinities? A review of fossil and recent Gypaetinae of both the Old and New World. *Contributions to Vertebrate Evolution*, 5:1-115.

Ritchey, K.A.
    1948. Lower Pliocene horses from Black Hawk Ranch, Mount Diablo, California. *University of*

California Publication, Bulletin of the Department of Geological Sciences, 28:1-44.

Robertson, J.S., Jr.
1976. Latest Pliocene mammals from Haile XVA, Alachua County, Florida. Bulletin of the Florida State Museum, Biological Sciences, 20:111-186.

Russell, L.S.
1968. A new cetacean from the Oligocene Sooke Formation of Vancouver Island, British Columbia. Canadian Journal of Earth Sciences, 5:929-933.

Savage, D.E. and L.G. Barnes
1972. Miocene vertebrate geochronology of the west coast of North America. pages 125-145. In E.H. Stinemeyer, Proceedings of the Pacific Coast Miocene Biostratigraphic Symposium, Society of Economic Paleontology and Mineralogy, 364 pages.

Savage, D.E. and D.E. Russell
1983. Mammalian Paleofaunas of the World. 432 pages. Reading, Massachusetts: Addison-Wesley Publishing Company.

Schultz, C.B. and C.H. Falkenbach
1940. Merycochoerinae, a new subfamily of oreodonts. Bulletin of the American Museum of Natural History, 77:213-306.

Schultz, C.B., and T.M. Stout
1948. Pleistocene mammals and terraces in the Great Plains. Bulletin of the Geological Society of America, 59:553-588.

Schultz, C.B., M. Schultz, and L.D. Martin
1970. A new tribe of saber-toothed cats (Barbourfelini) from the Pliocene of North America. Bulletin of the University of Nebraska State Museum, 9:1-31.

Schultz, G.E.
   1977. The Ogallala Formation and its vertebrate fauna in the Texas and Oklahoma panhandles. In G.E. Schultz, Guidebook for field conference on late Cenozoic biostratigraphy of the Texas panhandle and adjacent Oklahoma. Department of Geology and Anthropology, West Texas State University Special Publication No. 1. 160 pages.

Shattuck, G.B.
   1904. Geological and paleontological relations, with a review of earlier investigations. pages xxxiii--clv. In W.B. Clark, G.B. Shattuck, and W.H. Dall, The Miocene deposits of Maryland. clv + 543 pages + cxxxv plates. Maryland Geological Survey [1963 reprint]. Baltimore: Johns Hopkins Press.

Sheppard, R.A. and A.J. Gude, III
   1972. Big Sandy Formation near Wickieup, Mohave County, Arizona. Contributions to Stratigraphy. United States Geological Survey Bulletin, 1354-C. 10 pages.

Short, L.
   1966. A new Pliocene stork from Nebraska. Smithsonian Miscellaneous Collections, 149:1-11.
   1969. A new genus and species of gooselike swan from the Pliocene of Nebraska. American Museum Novitates, 2369:1-7.
   1970. A new Anseriform genus and species from the Nebraska Pliocene. Auk, 87:537-543.

Shotwell, J.A.
   1955. An approach to the paleoecology of mammals. Ecology, 36:327-337.

1956. Hemphillian mammalian assemblages from northeastern Oregon. Bulletin of the Geological Society of America, 67:717-738.

1963. The Juntura Basin: studies in earth history and paleoecology. Transactions of the American Philosophical Society, new series, 53:5-77.

1970. Pliocene mammals of southeast Oregon and adjacent Idaho. Bulletin of the University of Oregon, Museum of Natural History, 17:1-103.

Shufeldt, R.W.

1892. In E.D. Cope, A contribution to the vertebrate paleontology of Texas. Proceedings of the American Philosophical Society, 30:123-131.

1913. Further studies of fossil birds with descriptions of new and extinct species. Bulletin of the American Museum of Natural History, 16:285-306.

1915. Fossil birds in the Marsh collection of Yale University. Transactions of the Connecticut Academy of Arts and Sciences, 19:1-110.

Simpson, G.G.

1930. Tertiary land mammals of Florida. Bulletin of the American Museum of Natural History, 59:149-211.

1932. Miocene land mammals from Florida. Bulletin of the Florida State Geological Survey, 10:11-41.

Skinner, M.F. and C.W. Hibbard

1972. Early Pleistocene and pre-glacial and glacial rocks and faunas of north-central Nebraska. Bulletin of the American Museum of Natural History, 148:1-148.

Skinner, M.F. and F.W. Johnson

1984. Tertiary stratigraphy and the Frick collection of fossil vertebrates from north-central

Nebraska. Bulletin of the American Museum of Natural History, 178:215-368, figures 1-42.

Skinner, M.F., S. Skinner and R. Gooris
    1968. Cenozoic rocks and faunas of Turtle Butte, southcentral South Dakota. Bulletin of the American Museum of Natural History, 138:379-436.
    1977. Stratigraphy and biostratigraphy of late Cenozoic deposits in central Sioux County, western Nebraska. Bulletin of the American Museum of Natural History, 158:263-370.

Steadman, D.W.
    1980. A review of the osteology and paleontology of Turkeys (Aves: Meleagridinae). Contributions in Science, Los Angeles County Museum of Natural History, 330:131-207.
    1981. A re-examination of Palaeostruthus hatcheri (Shufeldt), a late Miocene sparrow from Kansas. Journal of Vertebrate Paleontology, 1:171-173.

Steadman, D.W. and M. McKitrick
    1982. A Pliocene Bunting from Chihuahua, Mexico. Condor, 84:240-241.

Stone, W.
    1915. Shufeldt on fossil birds in the Marsh collection. Auk, 32:375-376.

Suthard, J.
    1966. Stratigraphy and paleontology in Fish Lake Valley, Esmeralda County, Nevada. 108 pages. Master's thesis, University of California, Riverside, California.

Tedford, R.H.
    1981. Mammalian biochronology of the late Cenozoic basins of New Mexico. Geological Society of America Bulletin. Part I, 92:1008-1022.

1982. Neogene stratigraphy of the northwestern Albuquerque Basin. New Mexico Geological Society Guidebook, 33rd Field Conference, Albuquerque Country, II:273-278.

Tedford, R.H. and D. Frailey
1976. Review of some Carnivora (Mammalia) from the Thomas Farm local fauna (Hemingfordian: Gilchrist County, Florida). American Museum Novitates, 2610:1-9.

Tedford, R.H. and M.E. Hunter
1984. Miocene marine-nonmarine correlations, Atlantic and Gulf Coast Plains, North America. Palaeogeography, Palaeoclimatology, and Palaeoecology, 47:129-151.

Tedford, R.H., T. Galusha, M. Skinner, B. Taylor, R. Fields, J. Macdonald, T. Patton, J. Rensberger, and D. Whistler
In press. Faunal succession and biochronology of the Arikareean through Hemphillian interval (late Oligocene through late Miocene Epochs), North America. Berkeley, University of California Press.

Tordoff, H.B.
1951. Osteology of Colinus hibbardi, a Pliocene Quail. Condor, 53:23-30.
1959. A condor from the upper Pliocene of Kansas. Condor, 61:338-343.

Voorhies, M.R.
1984. "Citellus kimballensis" Kent and "Propliophenacomys uptegrovensis" Martin, supposed Miocene rodents are Recent intrusives. Journal of Paleontology, 58:254-258.

Voorhies, M.R. and J.R. Thomasson
   1979.   Fossil grass anthoecia within Miocene rhinoceros skeletons: Diet in an extinct species. Science, 206:331-333.

Warter, S.
   1976.   A new osprey from the Miocene of California (Falconiformes: Pandionidae). Smithsonian Contribution to Paleobiology, 27:133-139.

Webb, S.D.
   1964.   The Alachua formation. Society of Vertebrate Paleontologists Guidebook 1964 Field Trip, Gainesville, Florida, pages 22-29.
   1966.   A relict species of the burrowing rodent, Mylagaulus, from the Pliocene of Florida. Journal of Mammalogy, 47:401-412.
   1969a. The Burge and Minnechaduze Clarendonian mammalian faunas of north-central Nebraska. University of California Publication, Bulletin of the Department of Geological Sciences, 78:1-191.
   1969b. The Pliocene Canidae of Florida. Bulletin of the Florida State Museum, Biological Sciences, 14:273-308.
   1974.   [Editor]. Pleistocene mammals of Florida. Gainesville: University of Florida Presses.

Webb, S.D., B.J. MacFadden and J.A. Baskin
   1981.   Geology and paleontology of the Love Bone Bed from the Late Miocene of Florida. American Journal of Science, 281:513-544.

Webb, S.D. and N. Tessman
   1968.   A Pliocene vertebrate fauna from low elevation in Manatee County, Florida. American Journal of Science, 266:777-811.

Webb, S.D. and M.O. Woodburne
    1964. The beginning of continental deposition in the Mount Diablo Area. Guidebook, Annual field trip of the Geological Society of Sacramento.

Wetmore, A.
    1923. Avian fossils from the Miocene and Pliocene of Nebraska. Bulletin of the American Museum of Natural History, 48:483-507.
    1924. Fossil birds from southeastern Arizona. Proceedings of the United States National Museum, 64(5):1-18.
    1926a. Descriptions of additional fossil birds from the Miocene of Nebraska. American Museum Novitates, 211:1-5.
    1926b. Descriptions of a fossil hawk from the Miocene of Nebraska. Annals of the Carnegie Museum, 16:403-408.
    1926c. Observations on fossil birds described from the Miocene of Maryland. Auk, 43:462-468.
    1926d. An additional record for the fossil hawk Urubitinga enecta. American Museum Novitates, 241:1-3.
    1928a. Additional specimens of fossil birds from the upper Tertiary deposits of Nebraska. American Museum Novitates, 302:1-5.
    1928b. The tibiotarsus of the fossil hawk Buteo typhoius. Condor, 30:149-150.
    1928c. The systematic position of the fossil bird Cyphornis magnus. Canadian Department of Mines, Geological Survey Bulletin, 49:1-4.
    1930a. Two fossil birds from the Miocene of Nebraska. Condor, 32:152-154.
    1930b. Fossil birds remains from the Temblor Formation near Bakersfield, California.

Proceedings of the California Academy of Sciences, Series 4, 19:85-93.

1933a. A fossil gallinaceous bird from the Lower Miocene of Nebraska. Condor, 35:64-65

1933b. Pliocene bird remains from Idaho. Smithsonian Miscellaneous Collections, 87:1-12.

1934. A fossil quail from Nebraska. Condor, 36:30.

1936. Two new species of hawks from the Miocene of Nebraska. Proceedings of the United States National Museum, 84(3003):73-78.

1937. The eared grebe and other birds from the Pliocene of Kansas. Condor, 39:40.

1938. A Miocene booby and other records from the Calvert Formation of Maryland. Proceedings of the United States National Museum, 85(3030):21-25.

1940. Fossil bird remains from Tertiary deposits in the United States. Journal of Morphology, 66:25-37.

1941. An unknown Loon from the Miocene fossil beds of Maryland. Auk, 58:567.

1943a. Remains of a swan from the Miocene of Arizona. Condor, 45:120.

1943b. Fossil birds from the Tertiary Deposits of Florida. New England Zoological Club, 32:59-68.

1943c. Two more fossil hawks from the Miocene of Nebraska. Condor, 45:229-231.

1944. Remains of birds from the Rexroad Fauna of the upper Pliocene of Kansas. University of Kansas Science Bulletin, 30:89-105.

1957. A fossil rail from the Pliocene of Arizona. Condor, 59:267-268.

1958. Miscellaneous notes on fossil birds. Smithsonian Miscellaneous Collections, 135:1-11.

Wetmore A. and H. Martin
1930. A fossil crane from the Pliocene of Kansas. Condor, 32:62-63.

Whistler, D.P.
1970. Stratigraphy and small fossil vertebrates of the Ricardo Formation, Kern County, California. 250 pages. Ph.D. Dissertation, University of California, Berkeley, California.

White, T.
1942. The lower Miocene mammal fauna of Florida. Bulletin of the Museum of Comparative Zoology, 92:1-49.

Williams, K.E., D. Nicol, and A.F. Randazzo
1977. The geology of the western part of Alachua County, Florida. Florida Geological Survey, Report of Investigations, 85:1-98.

Wilson, L.E.
1935. Miocene marine mammals from the Bakersfield region, California. Bulletin of the Peabody Museum of Natural History, Yale University, 4:1-143.

Wilson, R.L.
1968. Systematics and faunal analysis of a lower Pliocene vertebrate assemblage from Trego county, Kansas. Contribution of the Museum of Paleontology, University of Michigan, 22:75-126.

Wilson, R.W.
1960. Early Miocene rodents and insectivores from northeastern Colorado. University of Kansas Publication, Vertebrata No. 7.

Woodburne, M.O., B.J. MacFadden, and M.F. Skinner
    1981. The North American "*Hipparion*" datum implications for the Neogene of the Old World. *Geobios*, 14:493-524.

Zakrzewski, R.J.
    1969. The rodents from the Hagerman local fauna, upper Pliocene of Idaho. *Contributions of the Museum of Paleontology, University of Michigan*, 23:1-36.

Index

abavus, Presbychen, 13, 44, 53
Accipiter species, 16, 99, 106
Accipitridae, 15, 25, 26, 29, 30, 34, 35, 36, 39, 40, 41, 42, 44, 46, 49, 50, 56, 57, 58, 59, 63, 65, 67, 72, 74, 77, 78, 81, 83, 84, 90, 91, 92, 93, 94, 99, 103, 105, 106
Accipitriformes, 14
Actitis species, 19, 56
acuta, Anas, 13, 56, 105
Aechmophorus elasson, 9, 94, 101
Aethia rossmoori, 21, 53
 species, 52, 81
affinis, Ortalis, 16, 61
 Aythya, 14, 103
Agate Fossil Quarries, 25
ajax, Protocitta, 23, 97, 100
Alachua Fm., 55, 78, 80, 84
alba, Ardea, 12, 95
albeola, Bucephala, 14, 99
Alca species, 20, 53
Alcidae, 20, 38, 52, 53, 55, 65, 69, 70, 74, 76, 81, 82, 101
Alcodes ulnulus, 21, 53

ales, Buteo, 15, 25
alfrednewtoni, Pinguinus, 20, 76
Alle species, 20, 76
Almejas Fm., 68
americana, Fulica, 18, 105
 Grus, 18, 93, 94
Ammodon Beds, 28
Ammodramus hatcheri, 4, 23, 77
Anabernicula minuscula, 14, 87
 species, 83
Anas acuta, 13, 56, 105
 bunkeri, 13, 80, 86, 87, 94, 99
 clypeata, 13, 105
 crecca, 13, 69, 95
 greeni, 13, 46
 integra, 13, 29
 ogallalae, 13, 61
 platyrhynchos, 13, 94
 pullulans, 13, 47
 species, 56, 79, 83, 90, 94, 100, 106
Anatidae, 13, 29, 34, 36, 38, 39, 40, 41, 42, 44, 46, 47, 48, 51, 53, 56, 59, 60, 61, 62, 65, 66, 67, 69, 71, 72, 76, 79, 80, 83, 85, 86, 87, 89, 90, 91, 93, 94, 95,

96, 97, 98, 99, 100,
101, 102, 103, 105,
106
Anatinae, 40, 42, 51, 62,
72, 75, 83, 91, 97
anglica, Diomedea, 9, 64
Anhinga grandis, 11, 55,
67, 72, 79
  species, 64
  subvolans, 11, 35
Anhingidae, 11, 35, 40, 55,
64, 67, 72, 79
Anser pressus, 13, 94
  species, 83, 105
  thompsoni, 13, 89
Anseriformes, 13
Anserinae, 40, 41, 42, 51,
53, 56, 65, 66, 67,
71, 76, 83, 90, 91
Antelope Draw, 35
antiqua, Australca, 21, 82
  Tringa, 19, 104
anza, Meleagris, 17, 105
Aphelops Draw, 63
aprica, Hirundo, 23, 93
Aquila chrysaetos, 15, 105
  species, 65, 83
Aramornis longurio, 18, 34
  species, 4, 56
Archaeophasianus mioceanus,
17, 33
  roberti, 17, 33, 43
Ardea alba, 12, 95
  polkensis, 12, 64
  species, 42, 55

Ardeidae 12, 42, 55, 64,
79, 84, 94, 95, 98
Ardeiformes, 12, 90
Ardeola species, 55
  validipes, 12, 95
Arenaria species, 19, 56
Arikaree Group, 33, 34
Arizona, 69, 82, 87, 90,
91, 96, 107
Arnett, 63
Asbury Park Mbr., 28
Ash Hollow, 46
Ash Hollow Fm., 46, 50, 54,
57, 58, 62, 67, 71,
85
ashbyi, Heliadornis, 10, 37
Asio brevipes, 22, 94
  species, 93, 105
asio, Otus, 22, 92, 97
Astoria, 27
Astoria Fm., 27
atavus, Palaeastur, 15, 25
auritus, Phalacrocorax, 10,
90, 94, 103
Australca antiqua, 21, 82
  grandis, 21, 65
  species, 76
avita, Coturnicops, 17, 94
  Microsula, 11, 38, 76
Aythya affinus, 14, 103
  species, 65, 106
Aythyinae, 76

Baja California, 68
Balearica species, 18, 30, 58
Balearicinae, 4, 28, 39, 41, 42, 51, 52, 58, 65, 67
barnesi, *Puffinus*, 10, 52
Barstow, 36
Barstow Fm., 36
Bartramia umatilla, 19, 80
baryosteus, *Pliolymbus*, 9, 92, 93, 98
Batesland Fm., 29, 31
Bear Creek Quarry, 57
Beck Ranch, 87
Belleville Fm., 106
Benson, 87
bessomi, *Oxyura*, 14, 105
Big Island Fm., 82
Big Sandy Fm., 82
Big Spring Canyon, 46
Black Butte, 47, 74
Black Hawk Ranch, 47
Blanco, 88, 97
Blanco Fm., 88, 89
Bone Valley, 63
Bone Valley Fm., 63, 66, 77, 78
Boreortalis laesslei, 16, 35
   phengites, 16, 52
   pollicaris, 16, 29
   tantala, 16, 25
   tedfordi, 16, 36
Botaurus hibbardi, 12, 98

Boulder Quarry, 41
Box Butte Fm., 30, 32
Box T, 66
Brachyramphus pliocenus, 21, 101
   species, 69, 101
Branta esmeralda, 13, 51
   howardae, 13, 60
   species, 44, 46, 56, 83
Brantadorna downsi, 14, 105
brevipes, *Asio*, 22, 94
British Columbia, 27
Broadwater, 88
Broadway Fm., 88
brodkorbi, *Gavia*, 9, 52
   Pliopicus, 22, 61
   Promilio, 15, 35
   Uria, 21, 55
Brule Fm., 33
Bruneau Fm., 68
Bubo species, 22, 65, 99
Bucephala albeola, 14, 99
   fossilis, 14, 94, 105
   ossivallis, 14, 65
Buckhorn, 89
bunkeri, *Anas*, 13, 80, 86, 87, 94, 99
Burge, 48
Burge Mbr., 48
Burhinidae, 20, 34
Burhinus lucorum, 20, 34
Buteo ales, 15, 25
   conterminus, 15, 63
   contortus, 15, 41
   dananus, 15, 78

jamaicensis, 15, 84, 103
species, 25, 35, 65, 83, 84, 92, 99
typhoius, 15, 25, 41
Buteogallus enecta, 15, 34
Buteoninae, 29, 36

Calabasas, 37
calhouni, Puffinus, 10, 53
Calidris pacis, 19, 65
   penepusilla, 19, 65
   rayi, 19, 79
   species, 56, 65, 79, 84, 86
California, 26, 36, 37, 43, 47, 49, 50, 51, 53, 54, 59, 60, 61, 67, 69, 70, 74, 81, 100, 104
californica, Diomedea, 9, 43, 52
californiensis, Mancalla, 21, 70, 102
calobates, Rhegminornis, 4, 17, 35
Calvert, 37, 77
Calvert Fm., 29, 37, 38
Cambridge, 66
Campephilus dalquesti, 22, 87
Canada, 27
canadensis, Grus, 18, 52
Cañana Pilares, 48
Cap Rock Mbr., 58
Capistrano Beach, 67
Capistrano Fm., 67, 70, 75
Capitonidae, 22, 35
Carlin High Level Quarry, 67
Carmanah Point, 27
Carmanah Point Beds, 27
Castle Creek, 68
Catharacta species, 20, 76
Cedros Island, 68
cedrosensis, Mancalla, 21, 69, 74
Cepphus olsoni, 20, 81
   species, 52, 74
Cerorhinca dubia, 20, 55
   minor, 20, 69
   species, 53, 102
Chalk Hills Fm., 68
Chamita Fm., 60, 72
Charadriidae, 18, 83, 102, 105, 106
Charadriiformes, 18, 38, 42
Charadrius species, 83
   vociferus, 18, 105
charon, Pliogyps, 14, 56
Chesapeake Bay Fauna, 38
Chihuahua, 85
Chimney Rock, 33, 43
chrysaetos, Aquila, 15, 105
Ciconia maltha, 12, 94
   species, 56, 64, 80, 83
Ciconiidae, 32, 39, 43, 56, 64, 73, 74, 76, 79, 80, 83, 94
Circus species, 16, 83
Cita Canyon, 89

Claremont Mbr., 61
Clarendon, 49
Clarendon Fm., 49, 59
Clifton Country Club, 69
clypeata, Anas, 13, 105
Coffee Ranch, 69
Colaptes species, 22, 97, 99
Colinus hibbardi, 17, 92, 99
  species, 87
  suilium, 17, 95
Colorado, 31, 40, 42, 85
Columbidae, 21, 35, 76, 82, 84, 99
Columbiformes, 21
concinna, Gavia, 9, 64, 71, 102
conferta, Grus, 18, 48
connectens, Megapaloelodus, 19, 29, 36
conradi, Puffinus, 10, 37
conterminus, Buteo, 15, 63
contortus, Buteo, 15, 41
Conuropsis fratercula, 22, 34
cooki, Cyrtonyx, 16, 34
Coraciiformes, 22, 35
Cornell Dam Mbr., 39, 41
Corona del Mar, 69
Corvidae, 23, 40, 48, 70, 73, 77, 84, 88, 97, 99, 105
Corvus species, 23, 77, 84, 88

Coturnicops avita, 17, 94
  species, 83,
Cracidae, 16, 25, 29, 31, 35, 36, 52, 61
crataegensis, Probalearica, 18, 32
Crazy Locality, 48
crecca, Anas, 13, 69, 95
Creccoides osbornii, 17, 88
Crookston Bridge, 38
Crookston Bridge Mbr., 38, 39
Crookston Bridge Quarry, 39
Cuculidae, 22, 31, 76
Cuculiformes, 22
Cursoricoccyx geraldinae, 22, 31
Cyclorrhynchus species, 21, 76
Cygninae, 76
Cygnus species, 13, 83
Cyphornis magnus, 11, 27
Cyphornithidae (see Pelagornithidae)
Cyrtonyx cooki, 16, 34

dakota, Strix, 22, 29, 30
dakotensis, Neophrontops, 16, 47, 74
Dalles Fm., 79
dalquesti, Campephilus, 22, 87
dananus, Buteo, 15, 78
Del Gado Drive, 49

Dendrochen robusta, 13, 29
Dendrocygna eversa, 13, 87
   species, 56
Devils Gulch, 39
Devils Gulch Mbr., 39
Devils Gulch Quarry, 39
diatomicus, Puffinus, 10, 49, 54, 55
diegensis, Mancalla, 21, 74, 102
Diomedea anglica, 9, 64
   californica, 9, 43, 52
   milleri, 9, 43
   species, 27, 52, 53, 75, 81, 101
Diomedeidae, 9, 27, 37, 43, 52, 53, 64, 75, 81, 101
discors, Podiceps, 9, 92, 93
Dissourodes milleri, 12, 39
dominica, Oxyura, 14
Dove Spring, 49, 50
downsi, Brantadorna, 14, 105
Drewsey Fm., 47, 73, 74
Driftwood Creek, 50
Dry Creek, 90
Dry Mountain, 90
dubia, Cerorhinca, 20, 55
Duncan, 91
Dunlap Camel Quarry, 28

East Clayton Quarry, 57
East Sand Quarry, 41
East Surface Quarry, 41
Eastover, 61
Echo Quarry, 41
Edson, 70
effera, Proictinia, 15, 25
Egelhof Quarry, 41
Egretta species, 55, 64, 94, 95, 98
   subfluvia, 12, 84
El Sereno, 50
elasson, Aechmophorus, 9, 94, 101
elegans, Rallus, 17, 94
elmorei, Larus, 20, 65
Emberizidae, 23, 77, 86, 88
emlongi, Mancalla, 21, 102
enecta, Buteogallus, 15, 34
epileus, Promilio, 15, 35
Eremochen russelli, 13, 47, 86
Ereunetes (see Calidris)
Erolia (see Calidris)
erythrorhynchos, Pelecanus, 12, 90
Esmeralda Fm., 50
esmeralda, Branta, 13, 51
Etchegoin, 70
Etchegoin Fm., 70, 71
Eudocimus species, 12, 64, 76, 92, 98, 100
eurius, Palaeostruthus, 4, 71
eversa, Dendrocygna, 13, 87

Falco species, 16, 74, 82, 92, 99
Falconidae, 16, 34, 41, 42, 59, 74, 82, 92, 99
Farmingdale, 28
farrandi, Jacana, 19, 56, 79
felthami, Puffinus, 10, 70
Feltz ("Feldt") Ranch, 71
femoralis, Phalacrocorax, 10, 37
fidens, Nycticorax, 12, 79
Fish Lake Valley, 50
fisheri, Pliogyps, 14, 99
flammeolus, Otus, 22, 97
Flat Iron Butte, 91
Fleming Fm., 45
Flint Hill, 29, 30, 31
Flint Hills North, 29
Florida, 30, 32, 35, 55, 63, 71, 72, 78, 80, 84, 95, 100, 103
floridanus, Phoenicopterus, 19, 65
  Promilio, 15, 35
Foley Quarry, 30
fossilis, Bucephala, 14, 94, 105
Fox Canyon, 92
Fratercula species, 21, 76
fratercula, Conuropsis, 22, 34
Fraterculini, 52
Fringillidae, 23, 88, 105

Fulica americana, 18, 105
  hesterna, 18, 105
  infelix, 18, 47
Fulmarus hammeri, 10, 53
  miocaenus, 10, 43
  species, 37

galbreathi, Miocitta, 23, 40
Galliformes, 16, 39, 40, 41, 42, 48, 51
Gallinula kansarum, 18, 99
  species, 87, 94
Gallop Gulch Quarry, 57
gallopavo, Meleagris, 17, 103
gambeli, Lophortyx, 17, 105
Garner Bridge, 40
Gavia brodkorbi, 9, 52
  concinna, 9, 64, 71, 102
  howardae, 9, 101
  palaeodytes, 9, 64
  species, 37, 75, 81, 101
Gaviidae, 9, 37, 52, 64, 71, 75, 81, 101
Gaviiformes, 9
Gaviota niobrara, 20, 54
geraldinae, Cursoricoccyx, 22, 31
Gila Group, 89, 90, 91, 96, 107
gilmorei, Proictinia, 15, 77
Ginn Quarry, 30

Glareolidae, 4, 20, 25
Glenns Ferry Fm., 68, 91, 93, 96
Goleta Fm., 95
goletensis, Phalacrocorax, 10, 96
Goodnight Beds, 73
Grand View Fm., 102
Grand View, 93
grandis, Anhinga, 11, 55, 67, 72, 79
    Australca, 21, 65
Green Hills Fauna, 36
Green Valley Fm., 47
greeni, Anas, 13, 46
Greenside Quarry, 34
Gruidae, 4, 18, 28, 30, 32, 34, 36, 39, 41, 42, 45, 48, 51, 52, 56, 58, 65, 70, 76, 77, 83, 91, 93, 94, 96
Grus americana, 18, 93, 94
    canadensis, 18, 52
    conferta, 18, 48
    haydeni, 18, 83
    nannodes, 18, 70
    species, 56, 58, 76
guano, Sula, 11, 64
Guymon, 80

Haematopodidae, 4, 65, 76
Haematopus sulcatus, 4, 19, 65
Hagerman, 93
Haile VI, 71
Haile XIXA, 72
Haile XVA, 95
Haliaeetus species, 15, 65, 76
halieus, Pelecanus, 12, 94, 97
hammeri, Fulmarus, 10, 53
Hans Johnson Quarry, 62
Harrison Fm., 25
hatcheri, Ammodramus, 4, 23, 77
Hawthorn Fm., 30, 31, 35
haydeni, Grus, 18, 83
Heliadornis ashbyi, 10, 37
Hemphill Beds, 66, 69
Herndez School House, 72
hesterna, Fulica, 18, 105
hesternus, Micropalama, 19, 87
Heterochen pratensis, 13, 39
hibbardi, Botaurus, 12, 98
    Colinus, 17, 92, 99
    Olor, 13, 94
Higgins, 72
Hill Point, 73
Hilltop Quarry, 35
Himantopus species, 19, 83
Hirundinidae, 23, 93
Hirundo aprica, 23, 93
Hogtown Creek, 30
Hollow Horn Bear Quarry, 51
homalopteron, Pandion, 15, 44, 76

howardae, Branta, 13, 60
   Gavia, 9, 101
hubbsi, Oceanodroma, 10, 67
Humbug Quarry, 41
humeralis, Sula, 11, 101

idahensis, Phalacrocorax,
   10, 64, 68, 94, 97,
   103
Idaho, 68, 82, 91, 93, 96,
   102
inceptor, Puffinus, 10, 43
incertus, Palaealectoris,
   17, 25
incredibilis, Teratornis,
   14, 105
infelix, Fulica, 18, 47
insignis, Laterallus, 18,
   99
integra, Anas, 13, 29

J. Swayze Quarry, 73
Jacana farrandi, 19, 56, 79
Jacanidae, 4, 19, 34, 41,
   56, 79
Jacona Microfauna Quarry,
   59
jamaicensis, Buteo, 15, 84,
   103
Jenkins Quarry, 41
Jewett Sand Fm., 26
joaquinensis, Plotopterum,
   10, 26
John Day Fm., 33
Johnson Mbr., 63, 85

Junco species, 23, 88
Juntura Basin, 73
Juntura Fm., 47, 74

kanakoffi, Puffinus, 10,
   101
kansarum, Gallinula, 18, 99
Kansas, 61, 70, 73, 77, 92,
   98, 103, 106
Kat Channel, 62
Keim Fm., 102
kennelli, Phalacrocorax,
   10, 101
Kennesaw, 40
Kern River Divide, 81
Kern River Fm., 71, 81
kernensis, Sarcoramphus,
   14, 81
Kimball Fm., 66, 67
kimballensis, Proagrio-
   charis, 17, 67
Kirkwood Fm., 28

lacustris, Rallus, 17, 94,
   99
laesslei, Boreortalis, 16,
   35
Laguna Niguel, 51
lagunensis, Praemancalla,
   21, 53
languisti, Pliodytes, 9, 64
Lari, 36, 58
Laridae, 20, 54, 65, 76,
   84, 99, 101

Larus elmorei, 20, 65
  species, 76, 84, 101
Laterallus insignis, 18, 99
Laucomer Mbr., 52
Lawrence Canyon, 74
Lee Creek, 75
Leisure World, 53
leopoldi, Meleagris, 17, 90
leptopus, Phalacrocorax,
  10, 47, 74
limicola, Rallus, 17, 105
Limosa ossivallis, 18, 65
  species, 84
  vanrossemi, 18, 55
Lisco Mbr., 88, 89
Little Beaver A, 54
Lomita, 54
Lompoc, 54
lompocanus, Morus, 11, 52,
  53, 55
Long Island, 77
Long Quarry, 34
longirostris, Rallus, 17,
  94
longurio, Aramornis, 18, 34
Lophortyx gambeli, 17, 105
  shotwelli, 17, 80
Loup Fork, 77, 78
Love Bone Bed, 55
lovensis, Pandion, 15, 56
Lower Snake Creek, 40
loxostyla, Morus, 11, 28,
  29, 37
La Goleta, 95

Lucht Quarry, 48
lucorum, Burhinus, 20, 34

macer, Phalacrocorax, 11,
  94
Machairodus Quarry, 62
magna, Paranyroca, 14, 29
magnus, Cyphornis, 11, 27
  Morus, 11, 52
majusculus, Podilymbus, 9,
  94, 98, 104
maltha, Ciconia, 12, 94
Manatee County Dam, 78
Mancalla californiensis,
  21, 70, 102
  cedrosensis, 21, 69, 74
  diegensis, 21, 74, 102
  emlongi, 21, 102
  milleri, 21, 74, 102
  species, 74
Marsland Fm., 25, 33
Martin Canyon Beds, 31
Martin Canyon Quarry A, 30,
  31
Maryland, 37
Mascall Fm., 43
Matthews Wash, 96
mcclungi, Miocepphus, 20,
  38
McGehee, 78
McKay Reservoir, 79
media, Miosula, 11, 52, 55
megalopeza, Speotyto, 22,
  93, 94, 99

Megapaloelodus connectens,
    19, 29, 36
  opsigonus, 19, 47, 69
  species, 44, 49
Melanitta perspicillata,
    14, 105
  species, 101
Meleagridinae, 4, 49, 56,
    65, 69, 76, 90
Meleagris anza, 17, 105
  gallopavo, 17, 103
  leopoldi, 17, 90
  progenes, 17, 87, 99
  species, 62, 65, 89, 95,
    104
merganser, Mergus, 14, 103
Mergini, 101
Mergus merganser, 14, 103
  miscellus, 14, 38
Merritt Dam (Brown Co.), 57
Merritt Dam (Cherry Co.),
    57
Merritt Dam Mbr., 57, 62,
    71
Mesembrinibis species, 12,
    92, 99
Mexico, 68, 85, 95
Michoacan, 95
micraulax, Puffinus, 10, 31
Micropalama hesternus, 19,
    87
Microsula avita, 11, 38, 76
  species, 53
Mill Quarry, 41

milleri, Diomedea, 9, 43
  Dissourodes, 12, 39
  Mancalla, 21, 74, 102
Minnechaduza, 58
minor, Cerorhinca, 20, 69
minuscula, Anabernicula,
    14, 87
miocaenus, Fulmarus, 10, 43
  Archaeophasianus, 17, 33
Miocepphus mcclungi, 20, 38
  species, 38, 76
Miocitta galbreathi, 23, 40
Miohierax stocki, 15, 26
Miortyx teres, 16, 29
Miosula media, 11, 52, 55
  recentior, 11, 101
  species, 53
miscellus, Mergus, 14, 38
mitchelli, Puffinus, 10, 43
Mixson, 80
Modelo Fm., 37, 49, 50, 60
Momotidae, 22, 72
Monterey Fm., 51, 53, 54,
    61
Morus lompocanus, 11, 52,
    53, 55
  loxostyla, 11, 28, 29,
    37
  magnus, 11, 52
  peninsularis, 11, 64
  species, 37, 44, 68
  vagabundus, 11, 43
Mycteria species, 12, 56,
    79

nannodes, Grus, 18, 70
Nebraska, 25, 28, 30, 32,
   33, 34, 38, 39, 40,
   41, 48, 50, 52, 54,
   57, 58, 62, 63, 71,
   77, 85, 88, 102
Nenzel Quarry, 39
Neophrontops dakotensis,
   16, 47, 74
   ricardoensis, 16, 50
   slaughteri, 16, 94
   species, 83
   vallecitoensis, 16, 105
   vetustus, 16, 34
Nettion (see Anas)
Nevada, 50, 67, 98
New Jersey, 28
New Mexico, 44, 48, 58, 60,
   72, 89
New Surface Quarry, 41
nigricollis, Podiceps, 9,
   70, 104
Ninefoot Rapids, 96
niobrara, Gaviota, 20, 54
Norden Bridge, 41
Norden Bridge Quarry, 41
North Carolina, 75, 82
Nycticorax fidens, 12, 79
   species, 94
Nye, 32
Nye Fm., 32

Observation Quarry, 41
Oceanitidae, 10, 52, 67,
   101
Oceanodroma hubbsi, 10, 67
   species, 52, 101
Ocyplonessa shotwelli, 14,
   47
ogallalae, Anas, 13, 61
Ogallala Group, 46, 50, 51,
   54, 61, 63, 70, 72,
   73, 77, 80, 85
Oklahoma, 63, 80
Olcott Fm., 40
Olor hibbardi, 13, 94
olseni, Propelargus, 12, 32
olsoni, Cepphus, 20, 81
111 Ranch, 90
opsigonus, Megapaloelodus,
   19, 47, 68
Optima, 80
Oreana, 96
Oregon, 27, 32, 43, 47, 73,
   79, 90
orri, Osteodontornis, 11,
   43, 49, 52, 53, 61
Ortalis affinis, 16, 61
   species, 66
osbornii, Creccoides, 17,
   88
ossivallis, Bucephala, 14,
   65
   Limosa, 18, 65
Osteodontornis orri, 11,
   43, 49, 52, 53, 61
   species, 32
Otus asio, 22, 92, 97
   flammeolus, 22, 97

Oxyura bessomi, 14, 105
  dominica, 14
  species, 65, 86

pacis, Calidris, 19, 65
Palaealectoris incertus,
  17, 25
  species, 38
Palaeastur atavus, 15, 25
Palaelodidae (see Phoeni-
  copteridae)
Palaeoborus rosatus, 16, 29
  umbrosus, 16, 44
Palaeochenoides species, 75
palaeodytes, Gavia, 9, 64
Palaeonerpes shorti, 22, 50
Palaeoscinidae, 23, 61
Palaeoscinis turdirostris,
  23, 61
Palaeostruthus eurius, 4,
  71
Palaeosula stocktoni, 11,
  50, 54
paleohesperis, Uria, 21, 81
Palm Spring Fm., 104
Palo Duro Falls, 97
Palostralegus, 4
Panaca Fm., 98
Panaca Valley, Quarry #2,
  98
Pandion homalopteron, 15,
  44, 76
  lovensis, 15, 56
  species, 64, 76
Pandionidae, 15, 44, 56, 64

Parabuteo species, 15, 81
Paractiornis perpusillus,
  4, 20, 25
Paracygnus plattensis, 13,
  67
Paranyroca magna, 14, 29
Parulidae, 23, 35
parvus, Podiceps, 9, 102
Passeriformes, 23, 41, 42,
  57, 59, 71, 82, 95,
  98, 100, 105, 106
Passerina species, 23, 86
Paulina Creek, 43
Pawnee Creek Fm., 40, 42
Pawnee Quarry, 42
Pediohierax ramenta, 16,
  34, 41, 42
Pelagornithidae, 27, 32,
  38, 43, 49, 52, 53,
  61, 75
Pelecanidae, 12, 64, 75,
  90, 94, 97
Pelecaniformes, 10
Pelecanoididae, 10
Pelecanus erythrorhynchos,
  12, 90
  halieus, 12, 94, 97
  species, 64, 75
penepusilla, Calidris, 19,
  65
peninsularis, Morus, 11, 64
perpusillus, Paractiornis,
  4, 20, 25
perspicillata, Melanitta,
  14, 105

Phaethontidae, 10, 37
Phalacrocoracidae, 10, 37, 40, 47, 55, 64, 68, 72, 74, 75, 78, 79, 90, 94, 96, 97, 101, 103
Phalacrocorax auritus, 10, 90, 94, 103
   femoralis, 10, 37
   goletensis, 10, 96
   idahensis, 10, 64, 68, 94, 97, 103
   kennelli, 10, 101
   leptopus, 10, 47, 74
   macer, 11, 94
   species, 55, 72, 75, 79, 101
   wetmorei, 11, 64, 78
pharangites, Plegadis, 12, 56, 90, 98
Phasianidae, 16, 25, 29, 33, 34, 35, 38, 43, 49, 56, 57, 59, 62, 65, 67, 69, 73, 76, 80, 85, 87, 89, 90, 92, 95, 98, 99, 100, 103, 104, 105
phengites, Boreortalis, 16, 52
phillipsi, Rallus, 17, 83
Philomachus species, 19, 65
Phimosus species, 12, 99
Phoenicopteridae, 4, 19, 29, 36, 44, 47, 48, 49, 57, 65, 68, 76, 79, 83, 86, 91, 96
Phoenicopterus floridanus, 19, 65
   species, 57, 79, 83
   stocki, 19, 86
Phorusrhacidae, 18, 103
phosphata, Sula, 11, 64
Picidae, 22, 39, 50, 59, 61, 87, 97, 99, 105
Piciformes, 22, 41, 42
Pinguinus alfrednewtoni, 20, 76
   species, 65
Plataleidae, 12, 56, 64, 76, 90, 92, 94, 98, 100, 104
plattensis, Paracygnus, 13, 67
platyrhynchos, Anas, 13, 94
Plegadis pharangites, 12, 56, 90, 98
   species, 104
pliocenus, Brachyramphus, 21, 101
Pliodytes lanquisti, 9, 64
Pliogyps charon, 14, 56
   fisheri, 14, 99
Pliolymbus baryosteus, 9, 92, 93, 98
Pliopicus brodkorbi, 22, 61
Plotopteridae, 10, 26
Plotopterum joaquinensis, 10, 26

Pluvialis squatarola, 18, 106
Podiceps discors, 9, 92, 93
   nigricollis, 9, 70, 104
   parvus, 9, 102
   species, 64, 75, 79, 83, 87, 98, 101, 104
   subparvus, 9, 101
podiceps, Podilymbus, 9, 64, 95, 103
Podicipedidae, 9, 36, 55, 64, 70, 75, 79, 80, 83, 87, 92, 93, 95, 98, 100, 101, 103, 104
Podicipediformes, 9
Podilymbus majusculus, 9, 94, 98, 104
   podiceps, 9, 64, 95, 103
   species, 80, 83
pohli, Sula, 11, 60
Point of Entry, 63
Poison Ivy Quarry, 58
Pojoaque, 58
Pojoaque Bluffs, 59
Pojoaque Mbr., 58, 59
polkensis, Ardea, 12, 64
pollicaris, Boreortalis, 16, 29
Porter, 59
Pozo Creek, 81
Praemancalla lagunensis, 21, 53
   wetmorei, 21, 52, 81

pratensis, Heterochen, 13, 39
Pratt Quarry, 57
Pratt Slide Quarry, 57
prenticei, Rallus, 17, 94, 99
Presbychen abavus, 13, 44, 53
pressus, Anser, 13, 94
prior, Zenaida, 21, 84, 99
priscus, Puffinus, 10, 43, 53
Proagriocharis kimballensis, 17, 67
Probalearica crataegensis, 18, 32
Procellariidae, 10, 31, 37, 43, 49, 52, 53, 54, 55, 61, 64, 68, 70, 75, 101
Procellariiformes, 9
progenes, Meleagris, 17, 87, 99
Proictinia effera, 15, 25
   gilmorei, 15, 77
Promilio brodkorbi, 15, 35
   epileus, 15, 35
   floridanus, 15, 35
Propelargus olseni, 12, 32
Protocitta ajax, 23, 97, 100
Pseudodontornis species, 11, 27, 75
Pseudodontornithidae (see Pelagornithidae)

Psittacidae, 22, 34, 99
Psittaciformes, 22
Ptychoramphus tenius, 21, 102
Puffinus barnesi, 10, 52
    calhouni, 10, 53
    conradi, 10, 37
    diatomicus, 10, 49, 54, 55
    felthami, 10, 70
    inceptor, 10, 43
    kanakoffi, 10, 101
    micraulax, 10, 31
    mitchelli, 10, 43
    priscus, 10, 43, 53
    species, 37, 53, 61, 64, 68, 75, 101
    tedfordi, 10, 68
pullulans, *Anas*, 13, 47
Pungo River Fm., 77
Pyramid Hill, 26
Pyramid Hill Sand Mbr., 26

Querquedula (see *Anas*)

Railway Quarry, "A", 39
Rallidae, 17, 34, 41, 47, 51, 56, 59, 65, 70, 76, 79, 80, 83, 87, 88, 92, 93, 94, 99, 100, 104, 105
Ralliformes, 17
Rallus elegans-longirostris group, 17, 94
    lacustris, 17, 94, 99
limicola, 17, 105
phillipsi, 17, 83
prenticei, 17, 94, 99
species, 56, 65, 83, 104
ramenta, *Pediohierax*, 16, 34, 41, 42
Rattlesnake Gulch, 39
rayi, *Calidris*, 19, 79
recentior, *Miosula*, 11, 101
Recurvirostra species, 19, 44, 83
Recurvirostridae, 19, 44, 83
Red Valley Fill, 32
Red Valley Mbr., 30
Repetto Fm., 70
Republican River Beds, 77
Rexroad, 98
Rexroad Fm., 92, 93, 98, 100, 103
Rhegminornis calobates, 4, 17, 35
Ricardo, 50, 59
Ricardo Fm., 49, 59
ricardoensis, *Neophrontops*, 16, 50
Rincon, 85
Ringold Fm., 106
Rissa species, 20, 101
roberti, *Archaeophasianus*, 17, 33, 43
robusta, *Dendrochen*, 13, 29
Rollandia species, 9, 55, 79, 80

rosatus, Palaeoborus, 16, 29
rossmoori, Aethia, 21, 53
Round Mountain, 60
Round Mountain Quarry, 60
Round Mountain Silt, 43
Runningwater Fm., 28
russelli, Eremochen, 13, 47, 86

St. David Fm., 87
St. Petersburg Times Site, 100
San Diego, 100
San Diego Fm., 100
San Fernando Valley, 60
San Luis Rey River, 75, 81
San Mateo Fm., 74, 75, 81
San Pedro Breakwater, 54
Sand Canyon Beds, 41
Sand Draw, 102
Sandpoint, 102
Santa Fe River IB, 103
Sarcoramphus kernensis, 14, 81
Sawrock Canyon, 103
Schoetiger Quarry, 39
schultzi, Spizaetus, 15, 67
Scolopacidae, 18, 55, 56, 65, 70, 79, 80, 84, 86, 87, 99, 104
Scottsbluff, 33, 43
Seaboard, 32
Sharktooth Hill, 43
Sheep Creek, 34

Sheep Creek Fm., 30, 33, 34
Sherman Oaks, 49
shorti, Palaeonerpes, 22, 50
shotwelli, Lophortyx, 17, 80
  Ocyplonessa, 14, 47
Sinclair Draw Quarry, 41
Sisquoc Fm., 54
Skull Ridge, 44
Skull Ridge Mbr., 44, 59
slaughteri, Neophrontops, 16, 94
Snake Creek, 52
Snake Creek Fm., 52, 63, 85
Sooke Fm., 27
South Dakota, 29, 46, 51
Speotyto megalopeza, 22, 93, 94, 99
Spizaetus schultzi, 15, 67
squatarola, Pluvialis, 18, 106
Star Valley, 82
Stercorariidae, 20, 38, 76
Stercorarius species, 20, 76
Sterna species, 20, 99, 101
Sterninae, 58
stirtoni, Tympanuchus, 17, 29
stocki, Miohierax, 15, 26
  Phoenicopterus, 19, 86
stocktoni, Palaeosula, 11, 50, 54
Stonehouse Draw, 34

Strigidae, 22, 29, 59, 65, 92, 94, 97, 99, 105
Strigiformes, 22, 32, 34, 36, 42, 57
Strix dakota, 22, 29, 30
Studio City, 60
subfluvia, Egretta, 12, 84
subparvus, Podiceps, 9, 101
subvolans, Anhinga, 11, 35
suilium, Colinus, 17, 95
Sula guano, 11, 64
   humeralis, 11, 101
   phosphata, 11, 64
   pohli, 11, 60
   species, 60, 101
   universitatis, 11, 31
   willetti, 11, 49, 54, 55
sulcatus, Haematopus, 4, 19, 65
Sulidae, 11, 28, 31, 37, 43, 49, 50, 52, 53, 54, 55, 60, 64, 68, 74, 76, 101
Sunder Ridge, 36
Sweetwater Canyon, 71
Synthliboramphus species, 21, 102

Tachybaptus species, 9, 55, 79
Tadornini, 38, 51, 65
tantala, Boreortalis, 16, 25
Tarboro, 82
tedfordi, Boreortalis, 16, 36
   Puffinus, 10, 68
tenius, Ptychoramphus, 21, 102
Tepusquet Canyon, 61
Teratornis incredibilis, 14, 105
Teratornithidae, 14, 105
teres, Miortyx, 16, 29
Tesuque Fm., 44, 58
Texas, 45, 49, 59, 66, 69, 72, 73, 87, 88, 89, 97
Thistle Quarry, 34
Thomas Farm, 33, 35
thompsoni, Anser, 13, 89
Thomson Quarry, 34
Threskiornithinae (see Plataleidae)
Tick Canyon Fm., 26
Titanis walleri, 18, 103
Torreya Fm., 32, 33
Tringa antiqua, 19, 104
   species, 84
Trinity River, 45
turdirostris, Palaeoscinis, 23, 61
Tympanuchus stirtoni, 17, 29
typhoius, Buteo, 15, 25, 41
Tytonidae, 22, 56

ulnulus, Alcodes, 21, 53
umatilla, Bartramia, 19, 80
umbrosus, Palaeoborus, 16, 44
universitatis, Sula, 11, 31
University Drive, 104
University of Florida Campus, 30
Upper Harrison Fm. (see Marsland Fm.)
Upper Steepside Quarry, 36
Uria brodkorbi, 21, 55
   paleohesperis, 21, 81
   species, 52

vagabundus, Morus, 11, 43
Valentine Fm., 38, 39, 40, 41, 46, 48
validipes, Ardeola, 12, 95
Vallecito Creek, 104
vallecitoensis, Neophrontops, 16, 105
Valmonte Mbr., 54
Vancouver Island, 27
vanrossemi, Limosa, 18, 55
Vasquez Canyon, 26
Version Quarry, 41
vetustus, Neophrontops, 16, 34
Virginia, 37, 61
vociferus, Charadrius, 18, 105

Vulturidae, 14, 44, 56, 70, 73, 81, 99

Wakeeney, 61
walleri, Titanis, 18, 103
Washington, 106
Wasonaka yepomerae, 14, 86
Westmoreland State Park, 61
wetmorei, Phalacrocorax, 11, 64, 78
   Praemancalla, 21, 52, 81
White Bluffs, 106
White Operation Quarry, 44
White Rock, 106
Whitlock Oil Well, 107
Wickieup, 82
willetti, Sula, 11, 49, 54, 55
Withlacoochee River 4A, 84
Wray, 85

Xmas Channel, 62
XZ Bar, 85

Yepomera, 85
yepomerae, Wasonaka, 14, 86
Yorktown Fm., 75, 77, 82

Zenaida prior, 21, 84, 99
   species, 82
Zia Sand Fm., 48

Figures

Figure 1 (overleaf). Approximate correlation of the local faunas containing fossil birds with some relevant systems of chronology. After Blow (1969), Berggren and Van Couvering (1974), Ray (1976), Repenning and Tedford (1977), Armentrout (1981), Tedford (1981), Tedford and Hunter (1984), Berggren et al. (1985), Domning et al. (1986), and Tedford et al. (in press).

GLOBAL CHRONOSTRATIGRAPHIC UNITS			TIME IN M.Y.	GEO MAGNETIC TIME SCALE		NORTH AMERICAN LAND MAMMAL AGES			PACIFIC COAST STAGES		
SYSTEMS	SERIES	STANDARD AGES		EPOCHS	ANOMALY				BENTHIC FORAMS		MOLLUSCS
NEOGENE	PLEIST	CALABRIAN	2	(2)	old reu	BLANCAN		LATE			San Joaquin
	PLIOCENE LATE	PIACENZIAN	3	(3)	2A			E	DELMON- TIAN	Venturian Repettian	
	PLIOCENE EARLY	ZANCLEAN	4 – 5	(4)	3		EARLY	LATE			ETCHEGOIN
	MIOCENE LATE	MESSINIAN	6	5 6	3A	HEMPHILLIAN	EARLY	L	MOHNIAN		
		TURTONIAN	7 – 8 – 9	7 8 9 10	4 4A						JACA- LITOS
			10	11	5	CLAREN DONIAN	EARLY	L			MARGARITAN
	MIOCENE MIDDLE	SERRA- VALLIAN	11 – 12	C5 C5A	5A						
			13 – 14	AA AB AC AD		BARSTOVIAN	E LATE	LATE	LUIS- IAN		
		LANGHIAN	15 – 16	5B C5B			EARLY		RELI- ZIAN		TEMBLOR
	MIOCENE EARLY	BURDI- GALIAN	17 – 18 – 19	C5C C5D C5E	5C 5D 5E	HEMING FORDIAN	EARLY	LATE	"SAUCESIAN"		
			20 – 21	C6 C6A	6 6A	ARIKAREEAN	EARLY	LATE			VAQUEROS
		AQUITANIAN	22 – 23 – 24	AA C6B C6C	6B		L				

WEST COAST	ROCKY MTNS. AND SOUTHWEST	GREAT BASIN AND COLUMBIA PLATEAU	GREAT PLAINS	ATLANTIC COASTAL PLAIN	GULF COAST	PLANKTONIC ZONES	TIME IN M.Y.
							1
	Curtis R		Blanco			N22	2
San Diego	Dry Mtn.	Grand View	Cita Canyon			N21	
	Benson	Flatiron B.	Broadwater				3
		Hagerman	Rexroad			N19	4
			Saw Rock	Lee Creek	Upper Bone Valley		
Etchegoin	Wickieup	McKay				N18	5
Kern River			SNAKE CREEK FM. / ASH HOLLOW FM. / Hemp.			N17	6
							7
		Drewsey			Mixson McGehee		8
			Clarendon			N16	9
	Dove Spring	Black Butte			Love Bone Bed		10
	Ricardo		OGALLALA GROUP			N15	
		Fish Lake Valley				N14	11
			Valentine			N13	
						N12	12
Sharktooth Hill	Barstow		Olcott / Sand C			N11	13
				Calvert	Trinity River	N10	14
		Mascall				N9	15
			Sheep Creek			N8	16
			Flint Hill			N7	17
			Running water	Farmingdale	Thomas F Seaboard	N6	18
							19
			Agate Q			N5	20
Pyramid Hill							21
							22
						"N4"	23
							24

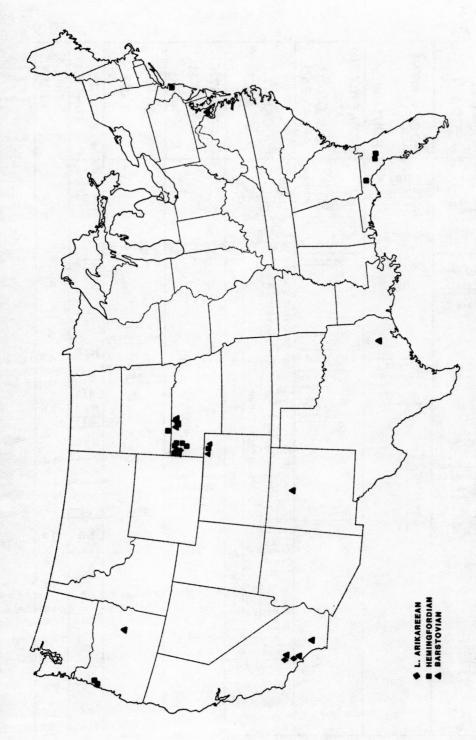

Figure 2. Geographic distribution of early Neogene (late Arikareean through Barstovian) localities in North America that have produced fossil birds.

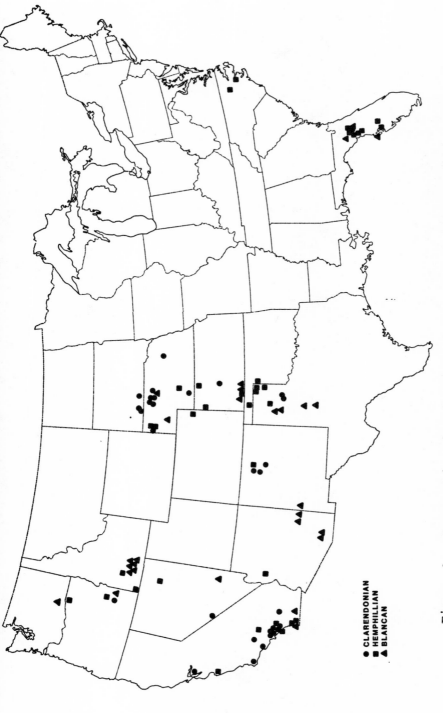

Figure 3. Geographic distribution of late Neogene (Clarendonian through Blancan) localities in North America that have produced fossil birds.